A freezing cold wi

hangs over a vat of hot water in the farmyard. The carcass will be scalded and scraped clean. A few neighbors have come to help and the women are already cleaning the chitterlings that will encase the sausage made from meat scraps.

Making sausage was my favorite part of the day. After the meat has been ground, then comes the seasoning: salt, red pepper, and sage from a nearby bush. The women mix it, then fry up a few "try pieces." "Needs more salt," one woman will say. "More pepper," says another. "You need a little more heat in there." The samples are passed around to get everyone's opinion—even a child's if I'm lucky because with so much going on, I've long since run off my lunch.

In *Shelby's Lady: The Hog Poems*, Shelby Stephenson has captured all the hard and often dirty work that goes into prepping and preserving a slaughtered pig before refrigerators were common. I remember salting the hams, drying the links of sausage or rendering the fat for lard. Shelby shows us again a way of life that has almost completely disappeared.

—Margaret Maron,
North Carolina Literary Hall of Fame

Shelby Stephenson's memories of time and place are so vivid you almost feel they are your own. He is both poet and singer, minstrel of hog-killing and connoisseur of barbecue, memoirist of affectionately named hogs, of family, work, and love. Shelby is a storyteller who can touch you like a sad Carter Family lyric, or an old gospel song, with many a twist of humor, as well as homage to dear kin such as his brother Paul. He finds poetry in forgotten phrases, in the minutiae of home.

—Robert Morgan,
author of *Dark Energy*

No poet I know of is as good as Shelby Stephenson when it comes to evoking the world through memory. If it's exclusively his world he welcomes us into, well, he does what every fine poet should do: he makes us feel at home by offering the occasional epiphanic observation couched in an endearing simplicity of style. And as always, there's a gentle presence in Shelby's poems, a crowd of the dead and gone who urge the poet into a wise exploration of the complex relationship of life to art.

—Stephen E. Smith,
author of *A Short Report on the Fire at Woolworths*

What other state can boast a song that dismisses witlings?

Being from North Carolina places Shelby Stephenson in the proud company of Thomas Wolfe, A. R. Ammons, Reynolds Price, Heather Ross Miller, Guy Owen, Fred Chappell, Jeffery Beam, Jim Wayne Miller, Ruth Moose, Robert Morgan, James Applewhite, Michael McFee, Robert West, and R. S. Gwynn—all of whom share the heritage of bracing shoreline and inspiring mountains, with a Piedmont in between where tobacco warehouses and cotton mills grow and furniture factories flourish, as well as almost as many pigs as people.

To be born in North Carolina in June of 1938—as Shelby and I both were—makes us the inheritors of unique protocols of landscape, history, religion, language, humor, literature, music, and—best of all—cuisine; and I envy Shelby's admirable power of embracing all these heritages in a powerfully authentic idiom that few can match.

—William Harmon,
J. G. Hanes Professor Emeritus in
the Humanities at UNC-Chapel Hill

I will drive far, through tumultuous and icy weather, to hear Shelby read from the rollicking barbecue poems herein. Don't mistake "country," as in hog killing and smoke houses, aunts named Vurtle and Auriba and remedies such as chitlin grease to cure coughs, for lack of brilliance. Who but Shelby could conjure this: "Harvey held one nose-hole / and let fly a greenish praying mantis of a stream." *Shelby's Lady* is pure Shelby. Pure joy.

—Dannye Romine Powell,
author of *In the Sunroom with Raymond Carver*

These well-crafted poems are what a reader would hope for from a master poet like former North Carolina Poet Laureate Shelby Stephenson, and yet, the poems in *Shelby's Lady: The Hog Poems* offer even more. One quickly remembers Stephenson is also a musician, and these poems sing with meter, rhymes/slant rhymes and rhythm, and the poems call and respond to one another. "Forget the Past," the first poem, is part road map for the book and part a (re)calling into being, a spell of the past cast into this present moment for the reader to experience. These poems ache with remembrance, shimmer with possibilities, and somehow teach us not only about rural life in Stephenson's North Carolina childhood, but also the larger truths of human connections to one another, to the land, and to the animals that live and work alongside us. The Duroc may be *Shelby's Lady*, but we get to visit her anytime at 985 Sanders Road near Benson within these pages.

—Malaika King Albrecht,
editor of *Redheaded Stepchild*
and inaugural Heart of Pamlico Poet Laureate

Yes, there is a music to life! In his Gershwin-style of storytelling, Stephenson sees and hears the melodies of nature that transcend this Earth as a sow feeds her piglets, "the sound/ is of a symphony, the boy inside the experience,/ the culture, not even hearing until time goes by/ what the sound builds." *Shelby's Lady: The Hog Poems*, in reverence to the pet pig of his youth and the cosmos, renders his overall good fortune of growing up during a time of "washboard" pastures, now turned into a monopoly board of houses surrounding his universe of Paul's Hill. With his unique humor, he recalls a guy who could sound exactly like a pig—then, "The rest is poetry." Stephenson's rhapsody, episodic with agony and joy, inspires with the beauty of pigs while celebrating barbecue in a world that "never tires of love."

—Hilda Downer,
author of *Bandana Creek* and *Sky Under the Roof*

The poems in *Shelby's Lady: The Hog Poems* do so much, and all at the same time. You get, of course, Shelby Stephenson's unmistakable voice, even if you've not heard it in person. It resides in Earth and in Heaven—sometimes in Hell. You get snitches, scenes, short stories, songs, and novels—sometimes in the same poem. And you get a kind of sadness and joy that could only come from inside the poem, inside the farm, the farm family, inside Shelby, inside the hog. This book sings, and wallows, and hollers, and then coos like a mourning dove. The poems take you to where they took place. Listen.

—Clyde Edgerton,
author of *Papadaddy's Book for New Fathers*

Shelby's Lady

The Hog Poems

Fr Ann once more Shelby 12-31-21

Shelby Stephenson

Shelby's Lady: The Hog Poems

Fernwood Press
Newberg, Oregon
www.fernwoodpress.com

Cover design: Mareesa Fawver Moss
Author photo: Juli Leonard, *The News & Observer*

Printed in the United States of America

ISBN 978-1-59498-072-5

Shelby Stephenson: Poet Laureate of North Carolina, 2015–2018. Recent books: *Paul's Hill: Homage to Whitman*; *Our World*; *Nin's Poem*; *Elegies for Small Game*, winner of Roanoke-Chowan Award; *Fiddledeedee*; *Family Matters: Homage to July, the Slave Girl*, winner of the Bellday Prize; *Maytle's World* (play); *Slavery and Freedom on Paul's Hill*. Recipient of the Distinguished Alumnus Achievement Award, 2015, Department of English, University of Wisconsin-Madison, he is Professor Emeritus, University of North Carolina-Pembroke, serving as editor of *Pembroke Magazine* from 1979 until his retirement in 2010. He lives at the homeplace on Paul's Hill, where he was born, near McGee's Crossroads, about ten miles northwest of Benson.

Also by Shelby Stephenson

Middle Creek Poems (1979)
Carolina Shout! (1985)
Persimmon Tree Carol (1990)
Finch's Mash (1990)
Plankhouse (1993)
Poor People (1998)
Greatest Hits: 1978-2000 (2002)
Possum (2004)
Family Matters: Homage to July, the Slave Girl (2008)
Playing Dead (2011)
Play My Music Anyhow (2013)
The Hunger of Freedom (2014)
Shub's Cooking (2014)
Steal Away (2014)
Eleven Poems (2015)
Maytle's World (Play) (2015)
Elegies for Small Game (2016)
Paul's Hill: Homage to Whitman (2017)
Our World (2018)
Nin's Poem (2018)
Slavery and Freedom on Paul's Hill (2019)

to the memory of my brother
Paul Stephenson, Jr. (1928-2008)
and always to Nin

Contents

Acknowledgments

Grateful acknowledgment is made to the editors of the following publications in which versions of these poems first appeared:

Contemporary Poetry of North Carolina (Anthology): "January Hog-Killing."
Heron Clan VI (Anthology): "Pink Pigdom"; "Confederate Soldier."
North Carolina's Roadside Eateries: "Visionary: The Starter of the Q."
Finch's Mash (Chapbook): "Piggybacking" first appeared as "Hogs."
Main Street Rag: "Thoughts of Where I Am From."
Middle Creek Poems (Chapbook): "When January Is Cold," first appeared in *Outerbridge.*
Negative Capability: "The Trench."
Negative Capability Online: "Chitlins."
North Carolina's 400 Years: Signs Along the Way (Anthology): "The Maw."
Plankhouse: "Remedies" (first published as "Old-Timey Remedies" in *The Panhandler*);
"Uncle John Woodall," first published as "Country Painter," *Appalachian Journal.*
Play My Music Anyhow (Chapbook): "The Vat: Farmers Depended On It," appeared first in
Wellspring; "The Orchard Boy" first published in *Cave Wall*; "Wind in the Woes,"
originally in *Kentucky Poetry Review.*
Wish You Were Here (Anthology): "If I Were a Razorback."
The following poems appeared in *Shub's Cooking*: "Pigs Feet";
"Stephenson, William Paul, Jr.—My Brother: The Q-Man";
"Tar Heel Barbecue" (first appeared in *Graze*);
"Barbecue" (first appeared in the anthology *I Let Go Of The Stars In My Hand*).

Forget the Past?

I

Various beds, some wallows. Sandy ones, too,
Like the mouth of the Buzzard Branch on Middle Creek,
The longing history of bob-wire, as we called it,

The hogs running out, for the "stock law" was in,
And the pastures were places to re-enter,
Swimmy-headed in the tops of mulberries

Overlooking the creek, overreaching the banks
Of myself and the dusk in the whippoorwill's
Call outside my bedroom window, the sweet transition

To darkness dipped in sumac and sassafras,
Willows, willows, weeping for me and you,
Trailing our names along without a hint of fear.

Tobacco and the bloom of biceps of girls
The sun shines as a freak of nature,
Erotica of sand between the toes,

Hoe-handles and jokes and shade,
The ends of rows in fields the South
Is known for, the short rest of my days

When all I knew was get to the end of the row
And turn around and go down another,
Another, another, and keep running

Like that without running out, dodging the high
Cotton where the sucker-pullers,
The Help, went to do their business.

It's when the fence is between the goose and you,
Right when tracks stop and you check your pants,
Ending the gander, if anyone's looking on.

II

So what'll I do, when you
Are old and gray, and I'm so blue,
What'll we do?

When you sat, cross-legged in the van,
I thought all heaven would transport
You and me to something no privilege

Could bear, as if a harbor of stillness
On a heather of nettles brushed rarer
Sunrises on a hedge of lilacs

My child's world wrought in separations
Up to now, territorial, without knowledge
My country could include us in its sunshine.

Now I know. Now that summer's
On its way and your face in mine
Becomes a sound of cool Missouri rain.

I was three, but I really maybe was three thousand
Trying to remember my first recall, sitting up
In bed at Uncle Reuben's and Aunt Mary's,

My ears so big I wanted to fly away,
As my father said, when he sang,
"Lonesome and sad, sitting by myself all day long."

That's all I know now, helped by a photo
Of Nin when she was a baby, sitting in her
Scant clothing on a bed a photographer

No doubt planned per her parents' request.
For me: a pile of years and no pictures.
Just cold memories of some old used-to-be.

III

The old people used to say, "disremember":
That boy? He got down on all fours like he
Was his own self a boar and he'd quote Bible

Verse and shout his name on street corners
As if he could wave his arms and the Spirit
Might find sorrow to stand with him for a moment

While he fumigated the boys in short-pants
At family reunions surviving
Deflections his mind today hazes

As if he'd know what to say sometime way
In the future when the pigs have gone to market
And home's a rib shack in the Heaven of Q.

IV

A prayer says: *God grant me the serenity,*
And the rest is wisdom, I suppose, deep in change
For accepting things as they are.

Alcoholics often cannot make step one
Of the twelve: I know a man who called from
A pay-phone, as he collapsed to a standstill,

The line dragging until a passer-by took him to AA
And struggled to release him to something other
Than that feeling of drowning in vain,

His claim to fame, this fabulous guitar-picker,
Freddy Fender told not to play chords
Before he finished introducing his song.

V

The guitar-picker kept his belly big once
He got sober: he loved to say, I been clean since
1982; that's almost forty years ago in the heat

Of coolness too, and I said, You have stood
Up for yourself, Julian, in a hedge
Of wild scallions and among the pricker-sleeves

Of jimson you have survived. I could hear the treble
And lingering notes from his Spanish electric guitar,
A slab of a sound under his elbow like a secret

He could turn on with his fingers and then
Say, "I just can't play under pressure, man,"
Talking that steady stand-in at the resthome, without blunders o

VI

Sowbelly of the South and hush little baby
In the cradle. Rock-nob where you latch
All air you lighten, crying

Out beyond memory, the mule, Gray, across
The field in her clear-keep, bob-a-longing leguminous hooves
On this Century Farm. Ultimate

Maytle, precise grabbles: if you could see
Her hands fanning in the dough so massive,
Gentle hums up-in-out-roll a dough-man?

Were you an artist to be in the white
Build-ups so clumpy and satisfying,
Your eyes on me, your fingers pinching that little dangle.

VII

To bring serenity into unity
Where culture pops and travel
Dodges in and out of leisure

With vans following the unanswerable
Line through life, he, alone, knew
What to do and how to do it:

He hitched his pants up, grabbed the wheel,
Amazed, told her to sit close as she could
To him and love him all the way

Into and out of accuracy, as she gave him
A smile, eyes, brows, the nose
And the length of her world to make their light.

VIII

And the just-shot hogs? Meanings that
Stick with the knife sharpening and shaping
The blood as the boy runs in and out

Among the hanging hogs and their dripping tongues,
The language a maze of promises
Harking back to bliss untranslatable,

The balloon full of dried peas,
The maw in the sun on the fence-post
Among the arching, aching sun

Bold for blood to spill over Paradise
Where children loiter around the gate,
The jar of whiskey *whshhhoooh—white lightning!*

IX

On this self-trip through Paul's Hill
Shub the FFA boy more than the 4-H-er,
Comes to give up the farming of pigs.

He is glad he never named them,
Remembering the words "Shelby's Lady"
For the only hog he ever really

Thought he owned, since it was a "project"
Delivered with papers to him
In a hogbox he made in shop

As a Future Farmer of America,
Greenhander, a hot stillness of sorrow
For lack of closeness for any pig he could ever call his pet.

X

Deserted childhood, I envy every stone
I ever saw, the old-field cemeteries,
The spaces between the graves, too,

The silence of masonry new ground
Blooms the way sight shapes black stumps,
The slow chains of harness and muscling mules,

Their rippling pulls my years
Consigning to dawns and joys
Of Depression-glass the plow-point

Might uncover in sandstone.
The little chores of equilibrium brim
My hair reddening like strawberries.

The Hog Trot

—adapted from the old song, "I Got a Pig"

I got a pig at home in a pen
Corn to feed him on
All I need's a pretty little girl
To feed him when I'm gone

Going up on a mountain
To sow a little cane
Raise a rip all night long
Little Liza Jane

Bare the soul that loves the biscuits
Baked so good and brown
Turn on the barnyard light
Shelby's Lady's running round

Shelby's Lady as 4-H Project

My 4-H Livestock Record Book was divided
Into parts: *Purpose*—(a seeming splurge),
Requirements, General Information (a hedge),
Feed-Record, Pasture-Record, Animals Sold (to be decided).

Its insides, mostly empty, the *Show-Record* with rue
My heart left blank, along with *Judging Contests* (pig-tassel
Marking *Other Expenses*) and a twitchy tail,
Other Income, example,

Financial Summary listed in my hand—
A $20.00 purchase price: *Summary*: I took steps
To "select this project because Daddy" did not resist
Allowing me "good pasture, clean water," some sassafras—
"A good place to breed my sow"; she could lick her lips.

Financial Summary? My sow I see on a float—
The year is 1951: Hank and Lefty crowd
The jukeboxes, my gilt's value always in doubt,
Including her weight of twenty pounds. Size? A shoat.

I had "a place to keep my pigs"—and shout—
"I can take care of her"—why consider me
Lucky, plus, "Daddy's pasture?" No smarmy
Words, I thought, just "four acres, with a clean stream,"
This "stream of water running all the way"—my vocation,
This branch in lower Cow Mire, plus tall
Shade trees, too, "grass and weeds for the hogs and cows"
    to grunt and low,
Eating stuff and drinking "fresh, clean water."
    What a vacation!

I signed the *Record Book*, "upon my honor," no rage,
At all, to say the least, my signature a trail
For cursive. My age I listed with a tail
Also, as "12," without fanfare or image.
The county agent's signature is missing; the sun
Still coming into my mind like a prayer: *I Pledge*
    (I hear Piglet snore),
My head to clearer thinking—
I am—*My Heart to greater loyalty* (Dorian Gray).

My Hands to larger service; and (my hat)—
I sunburn easily—*My Health* (various shoe-sizes)
To better living for (the miracle of sun rises)—
My Club, my Community, and my Country (all of that).

Three Little Pigs

1 *Shelby's Lady*

These were the days—
crossing their snouts,
Bacon and pork
chops, too.

Fatter they fattened
flabbier than any big fat sow
with one big boar
oinking, "I need you right now."

Shelby's Lady, meanwhile,
grunted in griddles of domesticana,
not foreclosing desire:
she wanted her back scratched

and the boy got his broken
tobacco stick and tied a rag on one end
and he dipped that rag and stick in a
bucket of cylinder oil

to kill what lice homed there
beyond his 4-H Project
the pedigree ruefully illustrated
with little boys and their pet pigs.

2 Norma Jean

The self-conscious walk
of a woman and her shoat, one quick
jerk of her wrist
and Norma Jean led the way

wriggling her tail
and I was walking
my Long Valley Jamie
Norwich, all shine and wear;

I said (we are on dunes at Sunset Beach)
"Hi"—and—"Is that a . . ."
and then—"My brother's got
a bar-be-cue . . .—

I mean—a restaurant . . ."—
my mind smaller than a
pig's foot
steaming in vinegar

and I went home, *The Fayetteville Observer*
in the box: on the Society Page,
there on a sofa, Norma Jean,
COMING OUT the headline thrummed.

3 *Sparkle*

The one Archie Ammons
writes about—"Hardweed Path Going"—
Sparkle keeps bleeding out,
O god of hogs,

pet pigs that spin high-hanging
gambrels into memory,
the poor family waiting
to be served Sparkle's liver or tail.

"Just a hog, boy,"
I can hear a neighbor farmer say,
just a fattening-sized hog
waiting for the blade.

Bleed on out into stars, the Milky Way,
the path, weeds, boots,
slops, the pet jo-reet,
and Kate the mule, glue, and Silver,

Hi-yo, most of all, the hoofing
sparkling oblivion of arms
reaching out into Sparkle's haslet,
the coming down, dribbling blood.

Fixing the Boars

The washboard-pasture rolls under thunder.
A boar tilts a dam toward a porker.
Daddy hollers for Roof Allen under
A mimosa, home-place: "Hold his hind-legs!"
Roof's sitting under the walnut, ready,
Gillette razor-blade, his fingers firmer,
Real excitement, I feel, the seasonal
Ritual, cutting a male pig's brash ass.

I don't know why I always hold dying,
Scaredy-pig squeals, as if sinews could hear
My brother Marshall sense his own body
Quiver as Roof's yellowing nicotine
Drifts toward Marshall's window, a faint mess
Of mountain oysters, Roof's face all ruddy.

Once More

I

The whole potato-patch field's a moat:
Every hog-path's a living water,
The cobs rolling, little logs afloat.

The big sow's with pigs: I had a few
Green collards for her from the marsh.
(We said "mash," like a brew.)

II

At other times, it appeared that retreating
Was out of the question, for rears
Twirled and blinked (little photos), dreams
Of home, a table set after school,

When I'd get off the yellow bus, raise the latch
To the kitchen door which slammed like a cannon
Sounding out for Durocs,
Especially Shelby's Lady, a cloud of much
Steam and smell of pigs-feet, roasted,
Though I preferred Mama's pork Q, crisp, browned well.

The kitchen table was homemade
And scrolled on the legs like a song,
The food piled smoke-high in long,
Narrow plates out of the polished times
Of Victoria Falls and awe-inspiring dormant moths.

The weedy path
Went down and curvy to the pen,
A fattening-cubicle built of outsides for pigs to share
For selling to buyers who sold to ones
Who sold to companies to make bacon.
And there they were,
Gathered together like an Old-Baptist church service,
The big one breathing, while nine pink piglets
Whispered "milk" around her tits.
The baby pigs seemed not constrained once
They clamped their snouts around
The pull-swelling, contented sweetness,
So good was the taste of the love-drink in that pasture.

When they turned the tits loose
They felt a forest of squeals,
A balcony weaning toward bacon
Like water flows over a dam and through.

Serenity

i

Serenity is the pasture of itself, the pigs
In coldish water snoozing another
Swelling wallow in the mud

Of bundles of fodder tied in the field,
Of longing earshots out of hearing,
Corn leaves crinkling the scene,

Shocking the sun in strands of turns,
The fodder-pullers not holding back the wave
Of bound shocks amplifying air and fading

Into wrestling matches among the strong boys,
Fair under the sky, the clouds
Of laughter the farm rides on.

ii

When I was a child down in Johnston County,
James Dickey I had not read to tell me
Of the lore of what might be in some bottle

In a back room museum in Atlanta: neither I
Nor the ewe-bushing gotcha of gothic etching
Could prepare a place to meet the tender meat

Of a sow in heat, say, *or* a sheep-child
In what mind the people who label such things
Naturally, as in the flea-wedding forever staged

In that little museum in Belhaven, North Carolina,
Outward, to now, to be the same tremble
As pig trotter stew in sweet black vinegar.

Shelby's Lady Chases My Brother Paul

There's a thing to remember about hogs
That happens if a sow's just farrowed pigs.
You better not think the hogs are playmates.
Actual story's true as rotting figs.

For a second, he split, my brother, Paul,
Put some corn in a bucket and whirling
As he swung himself in his arms and air
To pour shelled corn among a sow and pigs.

He said the mama sow collected him
In one showering, slobbering teeth-show
And she took after him, running him down
Over cobs and shucks, noses drooling goo.

I really wish I had been there to see.
Flourish spilled the grains—the trough; Paul, agog,
Eager, thrashing, hit the sow on her snout.
I can hear him say, "You are one dead hog!"

Solitude

I

There the hairy-legged boys lay,
Low down in the hayfield,
While the cocks strutted in the hay
As Goddess Tap-Tap sighed.
Richard with his bottles for refund
Of deposit for gas
And Skeets McDonald's itching bum
Inclined a bruise of grass;
A five-string picker nosed
Toward GTT, brushing a breast
And settling into rows
Of fuzz-balls velcroing the west.

II

They could have ganged up on a cow
Or sow in heat or sheep,
Or lit a mule's fart to warp wows
Incense could never leap
To thwart imagined stares, guffaws,
The pawing grins a dead
And endlessly pounding give-all
To the burdens which led
Them there to fumigate in rants
To bully flash's catch
From their hard and goat-bleating pants,
Off key, pitched, naked flesh.

III

Solitude lifts her curtain now
That she has changed her name
And grown scabs on rankism's roar;
Adolescence's time
Could not erase the slate to store
And keep her, the parting
Of the shades shuffling recall the more
The gas and fumes landing
On her sweating, singing belly,
The turns, out, in, entry,
As lovers, lost in the valley,
The moon and stars, voyeurs.

Once

I taught a hog to sing. I thought it sounded like poetry.
She was Shelby's Lady, pedigreed with ancestors named
    Joy and Ecstasy.
After a class or two I showed her off in 4-H.
She refused the right turns for oinks and slops.

Actually she got tired of seeing me deliver the swill.
She missed the flapping of my father's boots against the
    bends in his knees.

I think she did not want to throw off on the pone with tunes.
When I'd scratch her behind the ears, she grunted and wiggled
And drank thirstily from her trough.
I don't think my instruction was a waste.
I shoveled the offal from the outsides which floored her pen.

My father wanted to fatten her out.
I ceased to think of her as my student.
I dream of her sometimes.
My father loved her, said, "I think this is the best bacon
    I've ever eaten."

My childhood never got tired of loving her.
She was like bad poetry.
Now I come to my garden alone.
I do not think of Eve.
I think of her as a cat on a hot summer day.

If I were an animal I would like to be a human being.
The world never tires of love.
As a boy I yodeled in a changing voice.
Shelby's Lady knew. She could tell I was on to something.
She was a good musician. She could really imitate a hog.
I saw Joe Carter one time at The Fold.
He could impersonate a hog exactly.
The rest is poetry.
It happens when expectations become absences.

Gifts of Life

The first must be the violence of birth,
The beauty of that, out of one mother's dying
A little before abiding earth.

The second's got to be the routing
Childhood, without logos, scars, hoods,
The masked poem unwrit, biting

The lips of Pegasus whose mane
Ever dances in wearies for form
To put aside the page for hooves,

Otherwise the third would be give in,
Up, out, any drama unrequiting,
To eat the words on the tongue.

The fourth must be the hogs,
Shot, carved, dangling in gallows,
And one boy dancing as if buying

Jaw-breakers for the first hollow
In a tooth, no dentist handy to thirst
And say more than, "I'll pull it, no bother."

The fifth could be the WWII doc
Who'd seen meningitis in service
And forgive indulgences, saved my life.

The sixth is just a number, now,
Your life coming into mine, a guide
Or human post to be a flourish;

You shine for me like the crest on a mountainside.

Stories

I

The rain on Derek's Awning plops.
Splendor wrings the jay's call.
The snake slides slowly toward the wren's house
To get caught in the net for me to cull
And let go among the azaleas.
The fuzzy worms have clean-turned green.
Let the rain worship the metal and the grass.
Make the sun whirl reveries to pass
Not one given for Eternity.
The unknown knows no fear but itself,
Shut out from mid-August plenty.
My eyes drop inside my father's possum pelt.

II

I am still pointing a finger to a big oak
The worms ate or to hog-wire, mid-tree, my mother
Strung a clothesline on and hung out
Monday's wash so long ago I can't remember,
The car-porch, too, gone, except in my mind,
The driveway moved down the hill for better slumbers,
Quietude, as cars zoom after time.

Beyond the field across the road
Lived Uncle Reuben and Aunt Mary
And Joyce and Fred. My brother Paul's load
Of fun went crazy when "Thread" did a scary
Thing: tied his Schwinn to his dad's bumper
And let him pull him to Four Oaks.
A wonder he did not get killed, lumped
In a ditch with a sprocket in his throat.

There aren't many people around
Who remember when I was a boy.
Few pictures exist: my parents did not own a camera.
The graveyards flourish the joy
The hymns claim on Sundays
As the little churches keep up
Alongside years with little money
To offer old folks toward health during aging.

III

Some of my boyhood friends let politics
Woo their charms away from efforts
Articulation, like sticks,
Throws in a churning array of rings
To test capacity
Of short races the end
Begins on time to hold things
For what living lends.

I keep the strange-seeming
Possibilities—one man
Or one woman, an admirable beam
Her eye gleams without the also-ran
Of test and scores, lists, and bins
Of worn-out songs the sages have polished
For so long the glasses of etched scenes
Gather dust waiting to be relished.

IV

And I did not create Paul's Hill.
It is itself intact as the fox
That's getting smarter than the thrill
I feel from watching it in soft socks
Cross Sanders Road in sly
Reminder that the old parables
Still have cruxes of wry
Truth and many morals.

We don't play cards anymore:
Canasta, bridge, rook, rummy.
I guess I never liked board-
Games any more than Cricket, her tummy
Spread now in that chicken-pose
She's famous for; my pen flashes
Letters to sounds of vehicles over-dosing
On whizz, sling-sounds of races.

V

Once upon a time Sanders Road was not a name,
Suburbanized, I mean, with developers growing houses
In fields along each side of that dirt road, now a thoroughfare—
This hill, clay, once stuck with schoolbusses
Up and down like puppets in eyes
Of pupils (Cleveland's school-children were called),
Growing up to make fusses
Over how much money they might bank in time.

And many of those have not risen
From their places, though the graves
Sing on, the plots, runs for possums
On the prowl and mice searching for ways
To make their homes among
Our final resting places, while preachers
Adorn what calls they make for songs
To be sung for farmers, merchants, school-teachers.

VI

An old woman with a wagon of acorns
Brings me again out of memory
To sing refrain to Newton Grove and acres
Of pecan trees under her tenancy.
Take me to your room, dear Lessie, your bureau
Turning still toward the wall,
Your head a misery software
Does no good for you in your sorrow.

VII

My mother's cousins among the Johnsons!
The lines get longer with names:
Aldonia, Phylaster, Sophronia,
P. Jasper, Bright, Naomi, Nebraska, the same
Descendants who moved to Elevation where Marshall Perry
Hanged himself in that sorrowfully still-standing tree.
He lies at St. Mary's Grove, the mulberry,
His dying cannot free.

VIII

Into the fabric's being
I want to shout as far as I can to bring friends
To bear, be neighbors lost for words,
As I am: I want to hear breaths send
The paws of hounds scaling for voices
Of possum and raccoon hunters
And farmers who tote the lantern in choruses
For love of people, particularly mothers and O Baby-bunting.

I want to mine years most for times lost
With hunters and dogs on the race.
I cannot just toss
To win or lose the human-sway or dog's day
Remembered as some story or back-track
To finish a tale
That life is worthwhile after what the never-past has wrought.

Uncle Huel's Deep Do

The cold winters catch the farmer
        with pine-sides, nails, hammer.
It is fattening out time again.
        Uncle Huel is a whammer.

He secures the boards in the ground
        so the hogs cannot root.
The slops I pour smell up a sound.
        Uncle Huel gives a toot.

It's time to kill the fattest one.
        He yells, "Hand me that maul,
my boy, I'll show you how it's done."
        He puts both his feet all

scrunched between the two bottom boards,
        swings that maul in motion
which topples him over and towards
        the base, into, my lord,

the hog manure to disappear
        from my sight, my laughter
a cross between crowing roosters
        and a possum hissing.

In full dress-up my father comes
        running with a big rope.
He's singing, "Huel's stuck in the pen
        With the pigs—with some hope

We can pull him out—don't touch him,
he's covered head to toe
in big shoat-shit, just look at him!
He's turned into a plow."

The Dominicker hen dashes up
to pay respect and scratch
for a grain of corn on his shirt.
"Arf," barks Boogie to match

The awful trail Uncle Huel leaves
on his way to Cow Mire,
to wash his body's wads of weaves
from hogs I saw, I swear.

Pink Pigdom

What pig this is I point to say
Its squeal is the sound of rusty tin
The wind has brought to charm
Dead weight to momentum.
Just sheer lard and pork for the table.
Do not ponder the stuck hole
The neck reddens or the scald the trotters
Flourish without insult or remarks from bystanders.
It is another hog; gawking is not
A whimsy doodle inside a culture
Filled with yodels and frenzy,
Not one dull three-legged pun.
And when my brother opens the mouth
To place an apple in, just for fun,
I cannot acquire his frequency
Or feeling for exaggeration.
My view makes a difference now.
I love to taste and eat the jowl,
But I do not want to run out of meat.
I invent a self to hold a boy.
Association is the rule. I cannot weep
For time in the past and the blubbering boys
Riding their bicycles in and out of the hanged hogs,
Their bladder-balloons between their teeth.
Peach blossoms require some poetry:
The women are poking twigs up intestines again
To make chitterlings for Sunday dinner,
Which is really lunch, if you can call it that.

Semblances abound with analogies,
As a pig I never poke
Except to shush it in a chute
To take to market for the high dollar.
No longer personal, still property,
Like my neighbor's saying, "Just a hog, boy."

Teats or Tits

Hard for me now to say "teats."
"Tits" is the word: the cow's tits (I never could milk).
To see a sow lying in pasture, near a hog trough,
her pigs sucking: I have seen that: and the sound
is of a symphony, the boy inside the experience,
the culture, not even hearing until time goes by
what the sounds build, that noise beautiful
and inclusive, yet illusive as creation,
the sow snorting, grunting, the pigs attached
to the teats in suckling residue like mountains diving
up around the pink ears of the pigs feeding so tightly
yet loosely too, for the dull specter of whiteness
begins to run down over the mother's belly.

Tongue

So that's what the boy thought of
When he saw the hanged hog's tongue,
Red with blood, the gambrels, sun coming up
Golden and fuzzy like twine.
The Goddess Maytle got the washpot hot.
The chanters were dressed in OshKosh
With those little red zippers sewed into the bib.
That's where they carried their Beech-Nut
    or Brown & Williamson.
How they sang and chewed, since they say
It is here each hog shall be rolled into the vat,
Trace-chains of mules under the scalding flesh
Among the flapping wings of buzzards
And the slow-nose-to-face-on-posts kittens
Purring at the feet of my dear father.
Teach me, I said, to open the January cold with flames.
Carry me away on the shoulders and scarves of fire
Spreading out under the vat in wings,
The litard-knots hotter than a dollar pistol.
Get the knives sharp for Walter to go up from the throat
First, then down in one motion, the epic-emptying guts,
A blaze of years fanning
Over Cow Mire Branch, where the moon cheeses
And the hawthorn lights up its little lanterns
For dreams the boy will carry all his born days.
So he said he would love the starter-houses
Built in cul-de-sacs at his Rock Hole,
Where his swimming was not sacred, but scary as hell,
The other boy's hands, opening fingers on his head,

And holding him under the water which flowed
Into his eyeballs with big boughs white with foam,
A roof-grove in the deepening shadows
There where he would never leave.
And so the red tongue dried, the blood,
    still running toward the tree.

Thoughts of Where I Am From

(for John Yewell and Mimi Herman)

After the law and I failed each
other, spring '63, UNC,
a blind date got me a real peach-
job working for A. T. & T,
Long Lines, out of White Plains, New York,
buying easements and land over
seven states in the northeast.
December, '64, a rover,
seems to me, I took a leave
of absence (I am still on leave).
In '96, Nin and I, relieved,
came home again to never leave.
There is no doubt or stance about
that, just dignity memory
scores to keep us inseparable
all seasons, thanks to Paul's Hill.
It is located near 50-210
Fire Department & Rescue,
McGee's Crossroads, near Stephenson's
Barbeque & Nursery,
CVS Pharmacy (built
on land where my father and I
used to possum-hunt)
and, oh, we now have a Sheetz!
Smith's Nursery blooms downhill
from here, the plankhouse I was
born in—restored in the field.

I live in a brick ranch as
close as you can imagine
to Sanders Road, paved in '52.
On the mailbox to my lane
says 985—my heart renews.

A Former Drinker's Closed Ice-Box

Come fly with me, my little lass,
And keep the beer on ice
For anyone who asks to pass
The fiddler's sawing price.
It's not too high and its sound,
No swagger, no sloppy falls,
You just dance like a pig on its toes
And wiggle your collar bones.
I keep the beer way down in the cooler,
O mind your hair so sweet,
Don't let it tangle up and fool me
So I can't keep time with your feet.
Come reel around, my pretty little Miss,
We'll circle round Paul's Hill,
To old Shenandoah and your Mississippi
And Middle Creek—my beauty still!

The Music Cannot Escape

No, the music cannot escape me or any life, yours, my son,
daughter, family, extensions and confessors to justify
relations in and out of kneeling and rising to let words
syllable for plaques no one wants in the end, to put away
in backs of cabins in fate and faith for rest and dust and
rent: Archie Ammons's trophies, the Jones boy's guitar,
too (give it to a possum).

Giving up matters. Tex Ritter, Don Gibson, Hank Williams,
and many others in Hillbilly Heaven, appropriate
museums for prose and NoDoz; oh, how I care and give
consideration, quick, to primers I knew in my childhood,
the two books in the plankhouse the *Sears Catalogue*
and the Bible lying shut under dust on the top of the
chifferobe. Bring

on pleasure for the path to the pigs. Let purses open for Love
mingling with poetry-slams plus a gem from a favorite
lady to lay all matters to rest, without going to court, yes,
let the pulpit swell in fashions of courtiers, and count the
kings and queens to make sure the double corner stays
on the right on the checkerboard, lord, and vouchsafe the
deposit box for

money I never could count, let alone, word, and slop the
splurge of tears in the trough to dry when as a boy I
approached the trough timidly with my bucket of spill,
especially when the sow sidled up to me as if to say,
"How you do talk," and I bowed and nodded nobly to
her pigs (for I know what it is like to be chased), she, the
mistress, master, maker,

monarch and queen-bee mixing meadows with dandelions
and dandy dancers, singers and players, cardinals, too,
and more kings in this world, children, than you and
I can shake a stick at or lie down beside in waiting
for power to come on in North Carolina's General
Assembly, caps I give it, to act for mothers who teach for
not enough pay and for the

men who do not to fail to choose teaching little boys and
girls and big ones to startle ups and downs and skate on
slates out of respect for leaders that they may not lead us
into notoriety or to the notary public for a writ the pope
might not support. I blush for more objections to the
obvious. Let's part the clouds which look like heaven's
briars.

The Pasture's Quiet

It would not always be my father's lot:
he's in woods for a squirrel to loiter
into range; it is not yet quite sun-up.
My mother's already boiling water
in the kettle on the woodstove, the tug
of my body falling within the sun
patterning the gauze curtains in a snug
repeat in time: *Get up, you big-grown boy . . .*
When had I felt so heavy? Then one bright
array in the rapture, the pigs, all Durocs,
estranged from the pen now, snuffling the bit
of corn in the trough from last night's feed-up
in time oinked in the steady and certain
dip and glamour, as I wanted to end my chores
and skip to supper before darkness
scared me away every Sunday the more
I sank into Preacher Mills's sermons.
Praise-fests I transferred to the calf's halter,
yanking the yoke tight enough to get lost
in morning's sun-filled door. I drew water
from the lot-well. The fox loped by the door
to the feed-room. The meal in Lady's tub
I loved—that scent of molasses, the lore,
lure of sun ever shining, coming up.

The Poles

I'd say my father's glass was full when he said,
"Holly, let's drag this seine I made
Out of chicken-wire to the Peter Hole
On Middle Creek; boy, I bet we'll mop up the old
Horney-heads and horse-fish, the bottom feeders,
Creek's full of them; so just grab here—I'll
Lead; we'll be shitting in high cotton
In no time." Holly got sideways on top of
My father's "What? This bamboo
I wrapped with haybale wire so the fish won't go through
The holes. The poles have been lying on these tobacco-racks
Here, tier-poles?—for several years.
I gave Willie Stephenson a nice shoulder I dug out of the salt-box
For the poles. I think they'll be stronger than Old Bob,
    the ox I plowed as a boy."
I imagine a big channel-cat thrashing.
The seine sways like Sammy Kaye and Johnny Cash.
At the window, looking out over the sink,
Mama Maytle hums a hymn which does not
Change a whit the two men heading for the creek—
"I'll Fly Away," timing on the beat.
I take my place at the corner of the table (I'm the youngest child).
The horizon looks like a runny egg
Rounding up the sky.
Some figures appear beside
The man in the moon and dwell
In my lap—for a spell.
I think the two will come home with fish galore
And prove my father's strut more

Than his love of woods and a son-in-law.
They stomp the mud off their Ball
Band boots. In one swing of his right arm
He gets the corn-whiskey out of the barn.
Holly stands next to me; he looks like a not-quite whipped
warrior,
Out of some old chivalric story
I read in school: my father's tumbler fills.
Holly holds a shot-glass, as if he learned in college
To be polite and hug the glass the way
A possum might take a persimmon in its paws.
My man, though, turns his tankard up
And drinks it down in a gulp.
I hear no more about that seine until decades
Come and go; I am spreading pages
Of *Field & Stream* on the eatin-table.
"Whatever happened to that seine you and Holly were able
To drag to the Peter Hole?" Holly settles that at last:
"Shub, we wedged that thing between two sycamores where
I reckon it still is to this day."
Where did the stream go?
Into the snow where the tracks slow?
Then stop to form a little tune, a ditty?
What's a ruined spring? A wired sprung,
Whose mouth grows moths of shabby mosses
And bedding from inside old Martin guitars.
And what of the sunrise in the mist,
The undercurrent of territory a near miss for Cricket,
My lovely Norwich my shadow wants?
Strange how things escape me, like my first
Memory my line may toss and lasso

So I am sitting on the featherbed again
Or standing in the dark night
Howling to get back in bed, scared dream-head of Jesus,
My poise over the slop-jar
In front of the chifferobe, standing like
A bent coin pinging
A bucket for poles of longevity,
No scratch for the tests and scans
Of eyes on the bird that never falls,
Yet conjures the house-sparrow,
Spar-hawk, my father said,
That male with badge
My vision ever flutters,
Doors of bounce, hay-stuffed,
Near birches that bow their heads
In bracelets my pose rests
In thighs scraping me,
Our lives tossing and lassoing
The featherbeds again,
The squirrels in the feeder
Built like a little red schoolhouse,
A playground cannot leave.

If I Were a Razorback

If I were a razorback
And you were Shelby's Lady,
Would you tarry with me any way?
Would you farrow my babies?
If I were a wild hog
Would you still ride me,
Wiggling your tail through bogs,
Oinking up behind me?
Save my feral pig-dom
Across North America in Caroline.
Like Christopher Columbus, come,
Release the piglets into woodsy pine.
If I raked my hooves to bleed
Would you still care for me?
Tell me, Love, if you'd smear me on wood,
A door, like Huck, then escape with me?
If I were a farmer
With stock running free,
Would you allow your charmer
To root your tubers and seed?
If I were a razorback
And you were Shelby's Lady,
Would you tarry with me any way?
Would you have my babies?
Would you tarry any way?
Would you find our babies?

Pink Piglets

There's a chapel in the woods, mostly pine,
just a short walk down the hill lower there
in straw where the piglets lay in safeguard
with their mom all snuggled up like children
away from swords and children's blankets
and they wallowed in pink much younger
when the names of pigs they said were found.
The sow "found" the pigs to thrive on gain
of nipples pushing into mother's belly
boldly and more fully and cheerfully.
It was a time of love, of dwelling zones
which even now a precious foil
acts falsely to lie a shout away
from the bed where they lay without any
enemy in the world but God to keep
them safe from slaughter too soon for now
from stroke of blade or bullet between the eyes
to see other faces in the hurt-heart
where so fair they bedded for lords mornings
when the boy whistled to direct their play.

Hymn for Yearly Butchering

O holy post-Christmas when the blood began
To flow in cold January's heart
And the whiskey lit up the mouths of men,
When women dressed in frocks and sweaters, stripped guts
And sang of the Man after Adam, the One part
O holy after-Christmas when the blood began,
Our wonder that it ritually could be done
Already in the ordering of drink and food
While the whiskey lit up the mouths of men.
Walter's knife, sharp as a razor, caught the sun
At the hog's throat to split the hanging carcass into an arc.
O holy post-Christmas when the blood began,
The haslet, maw, the frilly white which ran
A bit down Walter's arm where he stood
As the whiskey lit up the mouths of men,
The women stripped intestines for stuffing sausage which sang
A smell the cat, meowing, circled for dribs,
O holy post-Christmas when the blood began,
And the whiskey lit up the mouths of men.

Duroc

CO-RUD would be a better name than *Shelby's Lady.*
Something about the latter in the mirror, mornings,
When I glance at the steam, wondering
If the day will be smooth and uncloudy,
Seeing Billy Wright putter on my account,
His John Deere pulling the hog-box I made in shop.
Foot-shuffling among the hedges, I was thirteen
And did not want to stand
Unable to respond to pedigree, reminding
Myself of dedication,
As well as the 4-H boy my situation
Numbed in name of puberty;
I felt my head aching, my ink-stained hand
Penning in the purchase price on the bill of sale.
Out on a limb of Carolina Ash at the pasture-gate,
I paid $20.00 to stand my ground,
Falling in love with the red, large-framed pig
That trotted up mornings for some corn, my pet,
Before the judge at the Swine Fest gave the ribbon
    to someone else.

The World as Duroc

I

Paul's Hill on a Sunday morning,
Eaves of the stable in sun-dew shining,
One by one my reveries goose-pimple
From heaven's roof, and I hug
My knees as if I am in a boat
Afloat on the wide Missouri or down on
My Middle Creek, longing for obedience
The titmouse sings for me,
The hog-parlor near as rainwater and urine,
My boy's life separate from the one
My dream highlights with buoyancy,
As a swim might quicken and loosen
A sail across the water the pasture shapes
Where the hogs let loose and twitch their tails
As if they might live forever.

II

Clearly the whole world could be a hog,
A perfect text of squeal
Where Judas wades into water
Up to his navel, looking for Jesus
Who lets John the Baptist hold a pig
On a leash in front of an abattoir
Where hogs are prepared for customers
Who receive the liver, feet, ears, noses
They want to make souse-meat out of,
The old way of killing hogs in cold January
As gone as air in a wave of eyes
Looking out at the zig-zagging life lived
As an invisible shadow or veil in beach-sand
Hurrying toward bright sunlight for a stone
Carved perfectly; the disciples gather
And stand up together for the slaughter.

Chester White

The Chester White breed of pig started years
Ago (1815-1818) right as the Romantic
Movement was stretching to peak with writers
Recreating intuitive feelings.
Shub surfaces among the pasturing lanes
To school himself in a wide channel
Of future farming at Cleveland High School,
All the time, his ag-teacher, growing dandelions.
Shub would even get a steer, Hereford, light
Red: Shub was not dedicated to beef.
He gave a fellow FFA student
A little sum to show that animal
So he, Shub, could travel on his senior
Trip by Trailways bus to New York City.
He borrowed a suitcase from brother Paul.
The hands of our chaperones spilled over.

Hampshire Poetics

I try to see the hog oink and not shriek
So that you really see one and stay on
The same page with me if a big king-snake
Crosses our path: I want the swine to reign
With ears alert; black body all its own,
The whitish band the middle shapes just so,
Covering the front legs; derivation,
Old English Breed rather than USA.
I recite northern England, Scotland, too,
Noting the well-muscled Catherine One
Gives a carcass-plus image for pork-view.
Praise the breed's capacity to unload
Extra longevity for the mother
That scores good tempers for the market-road.

A Porky Prospect

On the day of his leaving home, there were reams
Of poems left on Paul's Hill, plus some stump-struck
Chitlins which the boy remembered in a rut,
That trench he dug for the cat to purr
Around while waiting for one morsel to drop away
From Pearlene's hand, though she did not loiter.
She kept turning the entrails fast like a ride
Or a slide down Memory Lane for a sign,
Some prospect that a porky twinge might be a thing
To look for as entrance to a one-story house and rate
Whether a patient knows her big toe is being tucked in a throw.
That is a show the imagination bends in time.
You can push a hand away, and the little piggy needs frontage,
Some pasture to run in, all in a row
Down on the farm where the sassafras
Leaves in wind around the big oak a sense
Time keeps lounging, as if a squirrel were lazing
For the keeper to see the birds, all of them,
The titmouse, wren, hairy, red-bellied—
The cowbird, dove, chipping sparrow on a sash,
The window-pane, Carolina's bright
High-keyed calling, not kin to any shoat's
At all, although I have seen porkers on ridges,
Slobbering among cobs, unscathed,
The women around the trench right
On the edge of the offal with a limb like a lyre,
Peachy-ready; so go figure:
If you want to hear some piggerel, some rhymes,
Long enough for first grade class and supper,
Call up Tommy Tucker, Jack Horner, favorites—

The liquid eyes of Shelby's Lady
Squandering everything not herself,
As if, right now, she may be lying in her bed,
Snuggling for a ride with Razorback Roy
Rather than any old boar that swills up her mush
In her pasture on the hill, the pus
In her entrails pulsing her pussel-gutted squeal for pot-likker.
One time my mother
Walked out into our backyard. She had gone
To hang a haslet on a fence-post
And get some potatoes from the potato-bank.
Her apron wrinkled like a bird on a roof,
Her eyes moving scanners of light
Fixing their beams on running water,
Creek water, as a Duroc raised itself
From the hog-parlor and grabbed a bicycle,
Draped its front hooves across the handlebars,
And pedaled off into the tumbling world,
Merrily grunting and imitating birdsongs
Along the way, melodies of the wren, titmouse,
Cardinal, white-throated sparrow, the splatter
Of water in the shafts the sun jumps,
A quiet moment appearing out of nothing,
Different time, same place: dogwoods
Blooming April's story on the hill, my mother's heart
Settling into memory of my father, her husband,
For almost six decades, as the creek bank
He sits on whirls whitecaps over logs at the Rock Hole,
The little bubbles rolling through her hair,
Pure auburn tacking wind which wafts
Her apron up into a sail where she
Glides along the ditches on the shoulders of Sanders Road.

I Was a Blue Planet Chicken Once

A lookout, no beast, breast or grilled tofu,
but a sweet, natural sauce,
some cilantro, mozzarella
with pineapple and lean deli ham—
then I turned into a bird atop a blooming hardy
myrtle (I was thinking of Angier, North Carolina,
sanctuary to crepe myrtles
and bluebirds): I wanted to add
onions and bacon to some barbecue
and wince at the thought of pork,
a low cockpit, a crop-dusting baconer with wings
snorting toxaphene over cotton blooming
over the South, the yellow, waxy bolt-shoots
powdering my hair ungoggled as the manures
under my feet, dodging the white
parts, particularly, gerunds, I called that twang
I tasted in the feet of pullets
and in the rooster's cock-a-doodle-doo,
my goose-pimply reverie, my tree of rosemary
blobs, air, my post, tapping, topping with vegan cheese.

Serenade

I

"The redbud trees, Son, are blooming.
The catfish are jumping
And Middle Creek's waiting
For your uncle Walter to come pretty quick."
My father gets his lather out, a round
Wooden bowl, his hand on the brush's knob
As he shaves in front of the mirror,
Talking to himself again,
Another looking back
At the one in front of the lavatory,
Bare-chested, the face squinched to pace
His balance like a lever to see the world beyond Duroc,
The creek-longing evenings, the cool-screened lingering,
Keeping him nervous for what he takes for granted,
And so our world comes to order.
The Civil War's over! The slaves!
Praise the graveyards filled with connections
By that tall, dark-complected man
Who feared nothing all his natural days, though now and then
He had a thrill like the channel he caught
At The Mulberry Tree Hole, "Big as a young shoat."

II

Not far from where the cat runs with his lead-line,
In the family graveyard he hears the mourners
As his grandpa Manly gets lowered in the ground.
My father is ten all over again
And sits with a tree-crotch-view
Of the evening sun's rubble.

The Hog's Mine

"Grandpa Manly learned me a sense of history."
He says this to me as I have told many times.
"I have gutted and cleaned many a catfish."
My father says he had a pet pig one time
And his pa (my grandpa William)
Wanted to kill the pig for food.
My father said he cried.
I hurt to hear him say
His father said, "Son, you are mine, and the hog's mine."

A Tenant Wife's Vision

I remember this short woman sitting
In a straight-backed chair, her feet on a rung
One from the bottom one; she wore no shoes,
Her toes fragile as a purple martin's
On a nesting gourd: she stared, her fingers
Loving with a needle one striped work-sock farmers
Wore until the heel worked a hole that looked
Like a baby monkey's rear-end needing
Some attention. She looked beyond her scrubbed
Plank-floor, the biscuit-safe, the static
The Philco sent through me, wanting to say
Some mischievous thing as a boy could hear.
Her Roof was off at work for someone else
So that they could earn a place to keep and
A little garden grow in the black earth
Beside my father's home place: she, Roof, dwelled
On nothing deeper than trying to get by
From one day to the next: and now I know
Her mind—poem, song—was distinctly outdrawn
To me: I could feel her field in my way.

Uncle Shorty's Popped Pig

Of my father's three brothers, Shorty's one
Who comes nearly being real perfection:
When he helped his stroked, withered left side—he
Lumped along in a dragging kind of light
While he ate watermelon like an athlete
In competition with his pet pig, no sweat,
Or sat around, playing Setback, a sash
Tied around him in his chair; shoes shined, polished,
He knew his melon as one big game.
The black seeds he taught his pig to shine and
Roll each one which fell out of his fitted
Grin for a bantam to dart out and fetch,
And then, trust me, that pig went the farthest,
Over limits, popping right in two, wearing
The room, accommodations, wallowed space,
Entrails, a starry galaxy, soundless.
Please know this Hubert, my Shorty, his lead,
A sight together with his withered side,
His left hand, a deep white in my imagined
Aim for his pocket, now, the leafy hand.

Chester Honeycutt Revisited

My mother helped Chester show my white-faced
Hereford steer, my memory munching moos
On the Trailways to New York, senior trip.
I did not miss Shelby's Lady a bit,
Though hard facts now emerge: my big sore toe
The steer stepped on hackled a rash I got
From brushing its hide, the smell on the bus
Never leaving fumes to gather the walls
Of the shanty I was born in, my love
For home a high-flying bird, its mate lost,
Looking for color up the road, no rules
To find addresses in recent Lansing
Development where once hogs rooted pastures
Neighbors tended with certified farm-tools.

Hog-Box

I

My allegiance belonged to my Rambler, its rims
Glistening *I have learned the knack of pedaling*
And holding the handle-bars as I sped
And turned until I could sling a nasty at last,
Building a hog-box in shop blurred with spokes,
The way my hammer beat out *I believe*—things
About future farming, with *a faith born* with words, too,
As I would hum, while Mr. Horton put his bare
Face in mine, my reverie for any lace
The air could smell like guck until time frittered
Something in my groin the threads kept
Rubbery on tarred roads fresh as you
Came to be a decade later and now—your bed
High across the way as I call upon thee,
My angel, composite, variety
Caught up in silence, the object of my belief
In Shelby's Lady, my assemblage and belonging.

II

Then time took my care and the truth
That I did not mind anyhow?
My brother Paul became the farmer,
His longing dwelling in fields and growing hogs: he'd dole
Out money to tenants under trees,
The big elm at the Old Place a ruse
Still on one side of the images of my story.

I love that, too, the real kitchen smell,
The collards cooking on Duck's stove, as Earl,
Who had spells, knees up and down on his bicycle,
And then his flight over the handlebars
On the Claudy Fowler Hill, his helmet, intact, busted tires.
The sailing over the hood he lifted as his pedals
Wound and wheedled Percy's humming.
Turn around now to the shifting trail
Of the hog-box I made as project, the wheel
Of misfortune pouring dyes
Into my way of endless days
Helping me in my grit,
The slow gum of my loneliness.

III

Nothing rose to save me: beer, poetry, the rat,
That biggest one I called Wharf Rat, spotlit
By the barnyard bulb like OCuSOFT
Scrubbing my eyelids so I can bare *it,*
The center where the box gets lost: yet
The bike sings me back—my lariat.

The Sound Machine

A constant crashing of waves
On the counter sounds like snoring piglets:
I sleep without linking
Your long throw with mine, and then there
Was the Gibran and the lingering parting,
Wanderings of imperfect loveliness.
Our love could have been a long hard ride
Of a Hank Williams song: the pain
Of "I Can't Help It," my heart at your feet,
Unsteady from whiskey and beer,
Answering his songs with buckets of tears.
Or the thing could be the stress and separateness
Of two lovers leaving names we take on,
While standing under the clock where we first embraced.
There was a deepening sound, an amplifier tuning
Goodnight on hold and longing for water,
Salty-grained in sand and faraway white-caps.

Near the Hog Parlor

One time Mama was watching the "News with Dan Rather"
and he said something about puberty, and Mama
turned to me and said, "Son, what did he say?" I said,
"Puberty, Mama, you know that word?"
"No," she said, "I know Poverty. I did have a little wagon.
    It was red.
We had a little goat, and it was my horse."
She was standing at the sink in the kitchen. I was
about forty. She started talking about her father.
And she said she was six, waiting to walk to school
when her mother, name was Auriba, told Orron
to go to the hog parlor to check on his father
who went to feed the hogs. And she said her brother
Orron was twelve, and he came running back to the house
down there in Elevation: "It was forty-three days
after my family moved from around Newton Grove,"
she said: "Orron was crying, whining how his daddy was
dangling by a rope in a tree, a mulberry tree," she said.
I had never heard that story. The singing began
soon after. She started humming "Amazing Grace."
I never knew my grandmother Orby (no one could
pronounce the name *Auriba*). She died before I
came along in 1938. I think we
did have a happy life, didn't know any
better: my mother was one of nine children, two
dying in infancy of cholera (Charlie Emanuel
    and Parley Nathaniel).

You pick up what feathery song
in the air and go on. Grandmother Orby raised all the children,
never remarried. My mother married my father
when she was sixteen and he was twenty-one. They were
married for fifty-eight years. My life was like a
rose in a cleft, snug and surrounded by moss along ditches.
Forever seemed available and flowing
flavoring right up to waking up with
character and some clarity this morning:
now the hymn in the hum resides. My mother drifts
away in the shallows. I believe in the single fact
that happiness is in the first night love appears
under the stars, and purpose takes care of living
without ruling motion's a boredom
which will not sit still for long.
The world does not dissolve. It gets along
on its own until our hearts patter for Fear.

Right Around the Corner

She taught me what her six-year-old self taught her:
How the swinging father did not quit
Hanging in the mulberry she could not get right.
The legacy of that imagined show
Her brother actually saw echoed
Until he died, unable to loosen
His memory between his mother's block
Of desire and her husband, unable to listen
Anymore to pig-squeals as his world went black.

1

A mulberry limb swings 106 years ago
And returns to me, deep and wide, like a bone
Sharp as whatever sound he made inside his brow.
She's walking to school at Rehobeth.
She stops to strap a loosened sandal which slacks
Her skipping, this girl of six, free of that Sunday
Feeling church prompts when thoughts dollop
Fear even when she'd see a bunny.
Call her Maytle Samantha Johnson. Besides
A loyalty to my father for six decades of peace,
She comes home to me in several brides
I might have known, though the tone
Of her past and her instinct to survive the race
Drives me to leave my grandfather's grave alone.

2

She loved linoleum which stood out as rugs.
Her wood-cookstove was very white, black, and big—
A kettle on an eye which looked like a bug.
I would fill the reservoir, and she would give me an
  ice-cream cone
Filled with homemade delicacy she whipped up on the run.
The cow's butter shone in a shady corner from the sun.
I sat at the corner of the eatin-table in my high chair.
She never scolded me, even when my hands would whirr.

Standing at the Sink

Daylight: the rooster's crowing on tiptoe. Many little fluffy
biddies peep around the yard.
Mama Maytle's at the sink, humming into foaming rows of
dishes, her forehead home to a strand
of auburn, link to her mother, Auriba (Orby), the one whose
husband hanged himself in 1913 on a mulberry to the left
of the homestead the family moved to forty-three days
earlier.
I need some words to say how my mother lived from that
morning: she was six, waiting to go to Rehobeth School.
Her brother Orron was twelve and their, mother insisted
he go check on his
father who left in a hurry for the hog-parlor to see the new-
born piglets pink among the cobs.
The sow had farrowed the night before. Orron danced and
skipped to lore of found pigs.
He bobbed and shrieked at what he saw, his dad, Marshall
Perry Johnson, dangling on a rope he'd thrown over a
limb. Nothing could bring his father back, no one near
to wonder what
thoughts scoring the boy's breath cattybiarsoned balloons in
air that ugly-as-a-mog-owl February day, hope, gone,
and the space under the man in the mulberry, the future,
how and what to do.
Orron's thoughts run my own and get mixed up in Mama
Maytle's at the sink: images rain through the roof of the
ranch-home my mother finally got, after living the six-
year-old's dream

of wanting a father; she married her Paul when she was
sixteen and he was twenty-one, the two, for fifty-eight
years keeping their bond. Grandma Orby died before I
was born. A tall woman,
she was left with seven children, two more dying of *cholera
infantum* medicine could not
remedy in those days, when people were born to live and die,
whether neighbor or friend
or wifedaughtersistermother, washing and drying dishes,
cooking fried chicken, hoeing corn, sowing soda in the
middles of cotton or corn and cooking small game with
the instinct of
a Porsche mechanic. Before her death she lived beyond any
time, estranged as she was from her father who was not
there to help her mother raise those children.
So my mother raised us and sang in rhyme the songs she
knew, especially, "Will You Love Me When I Am Old?"
She hummed "Amazing Grace," the old reveries settling
laughter
and sharpening the clock. She lived until she could not, in the
suds, her dish-rag dangling like
a shoot on a mulberry she might jump up and tag before it's
too late to take off her apron
and hang it on the nail behind the pantry-door.

Poem for Squeezed Hands

She was six when her brother twice her age
Ran up from the hog-pen, the mulberry
Plenty strong enough to hold their dad's rage
On rope hanging a century ago.
The year: 1913: she scuffed the sand
To school in her Buster Browns, tarrying
Along in her brother's cry; fast, they ran
To mother: the little girl would marry
In ten years her dark-complected, one true
Love: he stood six-two, became my father.
I see her now with her brother, this boy—
Uncle Orron, who never took a bride,
Though he did love a Ferrell, first name, Doll,
Whose looks, my mother said, seemed certified.

The Varied Pursuits

for my brother Paul (1928-2008)

You suited up clothes:
striped ball suit was best.
It lasted through Old-Timers
ball banquets yearly
at Stephenson's Q
you started when hogs
you sold prompted
questions: "Why, what are
you going to do
with my hogs?" Guy said,
"Sell them, restaurants;
they make barbecue.
People come, greet, and eat,
charm plants through glass out-
side, trees, shrubs, plants."
So you, Paul, found one wax-myrtle,
Cow Mire Branch. You grew
fishing worms, too, night
crawlers you fed
first in sauce cans
from your Barbecue House.
And you took me into
your life, after you hit
balls, batting eighth, stance
sheathed in hands clawing
the bat to buttress
better ways to smack

drives to all fields,
delivering runs,
then running round your
centerfield position,
fielding grounders, catching
balls I thought
no one but you could field.
One coach, I won't name,
wanted you to palms up
your glove for balls coming
down in your place center
reckoned. I see you standing,
bones straight up, opposing
your way: ball hit you on
your chest which bore
threadmarks for months.
Our father knew
singing memory would
make out this song of
the ex-center fielder,
just as Shub writes
verse to walk morning
out into solitary silences,
moving pursuits to suit
up again, leaving fields
farmed, watermelons
grown among pigs to make
me swear I saw one piglet pop
in two: least I think
that hatched the tale
Shorty told: brighten

corners where you are!
Come, come home: churches
Jesus calls. What arm
is this reaching toward
heaven upon a couch,
while tall buildings flare
up and down Sunday morning,
now, all saints shouldering
history, bewildered,
as you, Paul, learning
brain cancer would waltz
across your mind four
weeks from day one
doctor Kemp spoke, his
laptop lit up to show up
that storm easing across
our eyes. You said,
I don't want operations.
That was that, no whys
Lining up days
blowing to be right
For blows which befall.

Romanticism

Since you said age makes a home where you've been
From call to call within circles of sin,
I must gauge my pain here at the homeplace
In plain sight of wheel-ruts and after-taste
Of persimmon, falls, memory's possum
Pock-marking bark, the pathway to Slopdom
Where as a boy I poured offal in troughs
For Duroc sows to grunt and grind what rough
Swill of table-scraps, garden-blessed and farmed
Right here on Paul's Hill, your wide-open arms
Too far away for me to face and bear,
Even though I love your raging everywhere,
Just out of reach of necessary light,
That delicate thief, wild, for your sweet blight.

Visionary: The Starter of the Q

He never wearied of what farming could do.
Tobacco? Gave it up for hogs, you see,
After taking some shoats he fattened out to Mr. Lee,
Down there below Benson. "He spit out his chew,"
Paul said, "right off, when I asked, 'What are you going to do
with my hogs?'"
"Sell them to a restaurant." "Mr. Lee shuffled the sandy rocks
Under his feet, as I unloaded my fourteen Durocs.
That was 1958. I went home, started this Q. There's the original
chopping block."
In the utter keep of growing things Paul had a touch.
He grew wax myrtles, nightcrawlers, melons and sold them
in the parking lot.
Neighbors said, "Well, you will go broke now, no one will come."
Paul kept playing the cash-register insomuch
As he knew music, standing there at his favorite spot.
Stephenson's Bar-B-Q he ran like he played guitar, in perfect time
his strums.

Stephenson, William Paul, Jr.—
My Brother: The Q-Man

There once was a man named Paul
With shoats he declared he sold
Whoops—he became ecstatic
When he heard a statistic
That he could barbecue his hogs
Said he to The Man he sold some to
What are you going to do with my sows—
To be understood without a blur
I'm going to sell them to a restaurateur
Unless you have a better idea
Paul Junior looked aslant
Truth struck a chord so blunt
To save his face
He put on his Mask
And instead of coughing he winced
As we travelled for every musical gig
Paul's humor was so big
He'd rub one hand over his balding and sway
The words You all just blow me away
As the crowd giggled in unison like one huge pig
Why William Byrd wrote about the hog.
He said North Carolinians eat so much swine
They fill themselves with grossest humors.
The Commissioner of Agriculture, James A. Graham,
One time said, in thirty-four days of campaigning
He ate barbecue twenty-six times.
I give up: barbecue's been here forever:
Ayden, N. C.—the Joneses, as early as 1830

Sold Q from a wagon; Ayden's known
Now as the "Collard Capitol of the World."
Before the Civil War
Ayden was Ottertown, I understand,
Because a man named Otter, Dennis Otter,
Would rather fight than switch from his corrupt ways.
"A-den," the town was called, for "den of corruption."
And when Dennis Otter left, forced to leave,
Ayden became known for its annual
Festival of the Collard.
Isn't that nice!
Now—think Melton's! Rocky Mount!
Melton's made a name for Q in the 1920s.
Now let's hear it for B's Barbecue, 761 B's Barbecue Road,
Greenville, NC 27858,
Whether weeeeeeeeeee want a little piggy moment or not!
How about the West?
Lexington BBQ #\1 or Luella's or Little Pigs in Asheville—
I just don't know: my thumb's rule?
Western North Carolina barbecue means *sliced* and
Eastern says *chopped.*
You can sense the merits of the chopped:
Go to Stephenson's Bar-B-Q.
Open the door there at the nandina.
There's the Chopping Block,
The original one—dated *1958*!

Barbecue

Is *it* a noun, adjective, or verb—as in *I like*
eastern North Carolina barbecue
Or *I attend Barbecue Presbyterian Church*
Or *You know that convicted murderer over there around Johnsonville,*
Why he got barbecued in the electric chair today in Raleigh.
(The thought of "shooting the juice," imagining hair singeing
on the chest,
turned me against the death penalty.)
Or *my brother Paul, started Stephenson's Barbecue in '58*
and his Q is vinegar-based, chopped, just the shoulders.
I knew Barbecue Hicks, third baseman for the Cleveland team
near Paul's Hill.
Short, stocky, Barbecue guarded his bag like—well—
Locals claim barbecue's the "other OTHER white meat"
(first is possum):
consider North Carolina the Barbecue Capitol of the nation.
I don't know: you see things for a very short time.
The coals or wood or chips get cold and then the smile's a pose
in a photo-shoot.
As you walk into Stephenson's Barbecue, after passing
the original chopping block displayed,
see the picture of my brother Paul, wearing horn-rimmed glasses,
his face sort of quenched up, not into an intended snout;
nevertheless, he looks like a pig, an advertisement beautiful
and lasting beyond his grave and ground-pepper
smell of vinegar, cooking pork, and sage.

Tar Heel Barbecue

for William Harmon

Western Q's good as Eastern—each is just a branch—
leg, shoulder, joint (sometimes)—
Give me barbecue in country or city where the pig's The Thing
that draws the diners
East is east and West is west (I'm talking *North* Carolina)
And the wrong one I can't try
Let's go where they keep on cooking
Those pigs and chickens—shoulders and wings
Noses and gizzards, livers, and fries,
Hush-puppies and (pork) skins—stew The Brunswick
Don't meet me at Sloppy Joe's
Amid chili's supremely doused bread
Take me where the vinegar-based sauce stings my tongue
And the mustard-tinged slaw invades my nose
I'll join you at Wilber's (Goldsboro—come again, Wilber's)
or B's (Greenville),
Holt Lake (Smithfield) or King's (Kinston) or
Stephenson's Bar-B-Q (Willow Springs)
or White Swan Q & Fried Chicken (five locations):
Smithfield, Benson, Pine Level, Wilsons Mill, Atlantic Beach
These and more—like The Redneck BBQ Lab, Benson,
Not far from Paul's Hill, the center of my universe—
East is east and West is west
For one can't go wrong on my Q's test
Imagine Somewhere Junction in the west: you can wear
your boots and jeans—
Buckskins, too—and conjure Love roving the prairie

Let's stop at the Barbecue Shack (Thomasville) for
shoulder (chopped & sliced) and
Tenderloin Biscuit heaven-sent, together with vinegar-based
slaw, hush puppies,
barbecued chicken, iced tea and soft drinks—no beer—and those
homemade biscuits like Mama used to make and serve hot
with cow's butter—yeah—
I'll love you in pigskin
Or shirts my Maytle made for me
But I'll love you more than anything else in the world
If I don't have to tote a gun or ride a bucking buckaroo
at Go Bananas
I want to suck on a rib and wing
My taste to announce the country music on the radio
turned down real low at
Lexington Barbecue Center, Inc.—
O roasted shoulder, minced or sliced, the smell of hickory or oak—
Give me eastern trimmings where the shoulder's thinner
Or western where pit-cooked's chopped or sliced—a winner
I don't need no French perfume to rock my room
I'm all yours at Stamey's (Greensboro)
"We cook our pork slowly over hickory or wood coals"—
Or Whitley's Luncheonette (Albemarle), a sandwich shop: you
can get a Q sandwich
chopped with no sauce—brings comfort to my stars
and O—while you're in that town
shout for Whispering Pines, hey, turn loose,
rave for that place—recommended by illustrator-artist Talmadge
Moose

Yes, and Red Pig at Locke Mill, Concord, just look
at all that history, plus Old Red Pig at two
locations, streets, Church and Corbin,
we cannot be redundant when savoring Q—
check out Gary's in China Grove
and Short Sugars Pit Bar-B-Q, South Scales Street, Reidsville
And I'm all yours whether a booth's cedar-wrapped or
    rustic-wiped with cloths
monogrammed just for us—our buttons—and blouses—skirts and
    squirts,
big and little, friends remembered or recalled, forgotten—
sauces, souse-meat, pig's ears, noses, vinegar-dressing, ketchup,
and that sweet iced-tea—
Take me to the original Little Richard's Bar-B-Q on
    Country Club Road in Winston-Salem!
Good golly, Miss Molly—have me—some fun tonight—
    Tar Heel Q's evermore consecrated!

Remedies

Put hog's foot oil and chitlin grease on a hot rag. Lay it
on the chest. That will stop the coughing. Rose had sore
eyes one time. They swelled way up. We put a green collard
leaf on them and they got well. They say the leaves
draw the fever. A piece of alum mixed with wild cherrytree
bark and whiskey is the best thing in the world. It
turns dark red. I got sick one time. Uncle John Woodall
gave me some. I came out of that bed like a pig.

Uncle John Woodall

Could grain a plank-wall, he said, so much like an actual oak
an old sow would lie down by it and perish to death
waiting for an acorn to fall.

When January Is Cold

In this ice-edged hour, this January of hog-killings, I see the whipped creak of trace-chains slipping under wrinkled snouts, pigs' lashes like drawn shade-tassels hanging from closed lids, know the running blood, the trembling jar of heads and ears on sleds mule-drawn to the barrel sliced in two bubbling with scalding water triple rainbows in the sun—I turn from the morning, and I believe in the first dying made in pleasure or pain and I feel the goneness, the sacrifices piling up in the fire growing around the lightwood knots under the vat and in the ice melting in dribs down hanging trees and I long for whole days of understanding the going-out lights, the washed-in-and-out of things in a January coming onto an old gallows tree when hogs are shot, cleaned and carved and salted in a box or hung up to the ceilings in smokehouses on nails and wires to cure, tongues dripping a language I hear.

Now

1

It is 2019, here on Paul's Hill: my hair's still red.
Grandfather Marshall Perry Johnson is still safe
In his place at St. Mary's Grove, in the graveyard,
Near rye-fields not yet risen where my mother,
His daughter, loved to stake the cow out and not miss
Being with the animals around her gathering.

2

"Why do you think he did that, a mess,"
She said, "There are easier ways than a mulberry."
Startled me. What could—I don't know, considering one
Less he made the family; Marshall and Orby were so close.
I mean: Henry M. Johnson married Edith Ann Allen in 1863.
Their son, Marshall Perry Johnson, was born in 1867.
He married Auriba Lee in 1892.

3

Marshall Perry Johnson killed himself on February 12, 1913.
He belonged to the Primitive Baptist Church for six years.

My sister, Maytle Rose Stephenson Hollingsworth left
Some notes based on a newspaper clipping of
        Marshall Perry Johnson's obituary.

He was a "kind husband and father," with many
        "good friends and neighbors."
They brought plenty of food to the house where
        he was laid out.

On Friday before he died, he saw a doctor who
    diagnosed him as having
"Nervous indigestion." The patient never got well
Again. Quiet in his last days, his eyes burdened to death.
One long morning, without warning, he died by suicide.

4

The newspaper article, Rose said, ended on a line
Of "great hope," how he was an "upright man"—reams
Could be written—oh how his burdens rolled away in linen
And how he would "see his Savior, face to face," there, then,
"Await no more headaches and longings," play some rook,
Perhaps be free from "this bitter world," the thin wind
Of "his sufferings," how great they were, like a whack
To hear "the doctor's skill in vain"; Marshall's hand
"Paid the debt we all must pay." He could button up
Freedom "from death and pain." He could not snap on
Life: "He is gone but not forgotten." Never an image will fade
Back to bring him here in a chair, though sweet rows
Of remembrances still linger in our minds for one sick man.

5

Around the tomb where holy days
Are laid, life is real and earnest, not weak.
The grave is not life's goal, I said, praise
That dust which scatters like a candlewick,
Yet does not speak for the soul which does not wreck.
Though he is gone from this shore, his spirit's bent
On troubles and pain no more. The next
Time we see him old acquaintances renew?

The point will rock and the rainbow steal into the soul
When the roll is called and dawn
Flows out like a scarf unfurling in a roll
Of times the children, especially, send sensations
Certainly for Marshall Perry Johnson who died
While his children grew and worked for bread.

6

Auriba (Orby), my grandmother, I never saw; a whir
Of memories base stories the weather
Brings to me, how she raised seven children right
As she could: Allen, Landon, Elam, not bad
Boys at all. Uncle Allen was in World War I, knew night
For sure, and Orron (never married): the bed
His father was laid out on could have overjoyed
Him, if that's the word, for Orron was twelve, his father's head
Aslant, hanging by that limb: Landon,
Like Elam, farmed. Charlie and Parley could not wear
Their twin-ship long. Infants, they died, emptied
By something like colic babies got then, separated
From proper medicines: the girls—oh how open
They were: Earle Washington, Vurtle, Maytle.

7

I thought of wrapping some soft lace
Around all the space, utterly rounded in bell-joys
Of branches in that mulberry: I lose my place
In the story. It leaves wallflowers as my source.
My uncles, the boys, were lively until they died.
My aunts never differentiated

My view of them as accurate as rye
Growing in a field of sophisticated
Wonder, with tips and green, fields, a ball
Of deep, planted rolls, with no evil
I can jam in a wad or dole
Out to anyone who wants to hear
About Marshall Perry Johnson. I never
Can get it straight: my mother named a son Marshall.

Take a Seat

Beside the little animals—the chipmunks,
squirrels, owls—until we see a line before ending
takes place and there is no other, the unknown

diving like the squirrel I saw this morning
sprint across the backyard, a universal
jump right where formation takes place
and we can stand up among our family

members and remember the good within.

January Hog-Killing

In water boiling every early January
just-shot hogs
turn over in chain-traces
horizontal across the vat
after the twenty-two's balling
the forehead *WHOMP* firm
shoatmeat falling dead on
putting the world out right between the eyes,
the jar passed round
into draughts,
hooch made in some radiator
abandoned for rotgut booze rising beads
bubbling down in crackling fire
ice-edged beams dripping hot running blood of hogs
gambreled puddling thawing dirt.

The Vat: The Farmers Depended on It

I

He slips in his Ball Band boots on the slick clayhill and slides in the slops spilling from his bucket on the path from the kitchen to the hog parlor pouring what's left in the wooden trough for the pigs and sows and one old boar, big cods banging.

Every hog that lumbered up from the woods was fattened out in a pen—that was the whole idea—for a cold-winter hogkilling day.

Pouring shelled corn among shoats Paul steps on a cob, loses his feet in the face of a rushing sow. . . the bucket across her snout and he is up over the pasture gate before he jumps, the sow bristling hair ready to run into weeds with a litter of pigs squealing and Paul sprawled in honeysuckle.

We dug a trench and stuffed it with lightwood knots. Paul poured kerosene in the trench under the vat and lit the fire. I would get as close as I could to the heat, waiting for the water to boil. I've seen Uncle Walter scald a 400-pound sow, turning it with chain-traces, and I'd scrape the hair off with jar-lids. Cigar clenched between his teeth my father would stride toward a hanging hog, tongue dripping red, and go up from the throat with his knife—and the lungs—what we called the "lights"—would spill a frilly cloud of whitish, red-spurting blood.

By crackling bread on the Home Comfort Range pigs-feet steam in vinegar. Daddy gets a hog's head from the safe and nobody knows but him the taste of eyes. I sit looking into rolled-up lids, eyes that might have stared at blue ribbons.

II

We stored the vat in the ordering-pit, a dug-out basement under the floor of the gradin-room. We hung cured tobacco there. Filling the vat with water helped bring the stems "in order." The extra dampness made the leaves soft for handling when we gathered upstairs in the fall to bundle the tobacco to get it ready for market.

More than once I raised the pit door, saw a moccasin or two slither on a tierpole, slip into the vat, imagined bubbling scalding water as it was for the hogs, smelled the singed skin, and saw the men lift the hamstrung, gambreled pigs on the gallows.

Piggybacking

I used to hear of the real, necessary juncture of nature and
rails, outsides, one side pinebark, other shaved. A boy
climbs into a pen, straddling urine's dirty canals, enough
to limber any bone. The years bring grit to grind with
mud. Hog tails hang down after the maul,

unchosen shoats and sows oinking noses through fencewire
on the Bob Higgins line. Men swill homemade brandy,
rake thick sleeves across their mouths cutting the sun's
slash on the tin barnroof—taking carcasses, pipes, livers,
hearts, lights into those bloody days, vatwater roaring
above the

flames, women ready to clean chitlins, galvanized tubs
emptybellied in ironweedstalks, gallows poking the sky.
The wrinkly end of the biggest gut opens: there's Dwarf
Tom Thumb, stuffed, eyeless, propped up against a
washboard: the sausagestuffer screwed to the eatin-table,
sour smell of intestinal

slop from stripped hog guts splattering the washbench,
foxhounds' backflung heads howling in the dogyard
at the boy rounding his bellowed cheeks to puff the
bladderballoon tied on a blackgum stick flitting in the
wingblur of gallowsbirds clapping among drying pig
blood, hunters knowing fresh meat

waits on the table when they remove their boots, put the
split floursack towel back on the 10-penny nail over the
washbasin.

The Poland China

When I was little I heard "Pole-N-China."
I thought if I dug deep enough I'd come up
For air and go all the way to China.

It was a long way in my mind
To cruise with pigs aboard a ship
Of sweet dreams shuttered up.

I saw the black hair, the six white patches.
If a hairy woodpecker were a pig
It might have wings all downy.

Big Bill weighed 2,552 pounds.
That's a lot of pork when you consider
He's got to hoof the hoof at high-pitch—

The squeal's the hog's bliss in the slops,
The rudimentary planted foot forward,
Without blue days, justice, corn and the mash

Again and again, piling on
The tongue-and-chew worthiness,
The round-up of a record, certainly

In the US—the highest in pork in pounds?
The Poland China stars naturally as originator.
It might have been a Polander by birth.

American Yorkshire

The breed was England's first; the US name
Domesticated from the white pig's splurge,
Enchanted pinkdom, flamingos could claim,
Ears not erect as the English Yorkshire's
Larger, cocked ears trembling, lying in rows,
Closed eyes, little pigs nursing at one time,
Filling hearts with more than the history
That Yorkshires came here in 1830.

They would populate real good everywhere,
The highest numbers—Midwest—Nebraska,
Illinois, Ohio, Indiana,
Iowa: the lean meat of boar and sow,
Back-fat, meaning more productivity
For its chow the industry might record.

The Trench

We called it the trench. And I do suppose
It could have been under skylights, a grove
Of chainey-trees, a slop-long, butchered, closed
Assembly, claustrophobia, from roof
Of sky to earth with organs a ceiling
For this offal, this edible weaving,
Female neighbors turning innards to match
These hot guts, one varied, largely small batch.

The food becomes gourmet's extravagant,
International cuisine, the loping
Deer and fox, mainly, my inhabitants,
Here, McGee's Crossroads where I am from—*Lord
How Mercy*—I smell that trench, the seven
Women—peach-tree limbs—stripping throw-away.

The Maw

Robert Dixon said he could gut a hog and have the haslet out in one minute and he went up from the throat with the knife and he did and someone put a tub under the stomach and it fell out, called it the maw. Nothing but dry corn in it. I got a peachtree limb and turned it wrongsideout.

We cut the maw free and rinsed it with hot water, swaying it back and forth, dropping one end into a trench. Then we placed it in a clean tub of water. We did the intestines the same way, emptying the tub, and the chickens would come and peck the corn.

Pigs Feet and Mama Said:

Wash
Cover in water
At boil, raise lid or else will boil over
& when they get boiled & water settles
Sprinkle as much salt as needed
Cook until tender (stick fork in them)—

Use lard for browning—for frying
Pigs feet—& put vinegar in pan
(cover pan) & let steam

The whole process becomes a lair of pigflesh
and vinegar, like bristles singeing where
hair was, the sty another bed after the
first farrowing forth of the sow.

Here are these feet crowded in my eye.
They are huddled together like pigs.

The toes could belong to shoats.
(A shoat's age is under one year.)
Idle persons are not worthless, necessarily, for
they see that the piglets are weaned.

One of the beautiful mysteries
Is to find the bed where the sow "found her pigs."
Pink pigs: all the way home from the woods
I would cry *weeeeeeeeeeeeeeee—*
I was happy.
My father would swill my thrill.

I wanted to be a farmer.
That's the wider current here, me—going to
        Beaver Dam, parting the limbs and
        leaves in the chapel to see the babies
        lying there, snuffling to find the tits.

I don't know how a little pig "had roast beef."
I have piggybacked many rhymes, running from
        the Nursery Books with a bun in my hand.
And I have seen in my living room my feelings turn on me.
I felt like a pig in a parlor.

I have poked a pig with a stick.
Is a pig happy in shit?
Seems so for a little while.

I imagine the feet in the pan belonging to a pig
        that took a sticker in its throat (I have
        seen that) the butcher knife going up from the goozle.

Big hogs show more of a scene.
I won't forget the haslet falling
Free into The Man's left hand.

Hogkilling did not command attention
        though the pitch and warp of the event
        left me high enough from the earth to
        receive the snoutblood—squeals—
        it all was as if an accident came about to
        circumstances—standing.

Chitlins

I

Snake-stealth. A bunch of wrinkles
Bulging with pig-shit on a limb,
        On a washbench, in a yard swept with brush-brooms.
Women, sweatered. Hedges shadowing sun.
Chimney. Plankhouse: one call, bracing hearts,
Mine, alone, in poetry loneliness

Sentimentalizes as buzzards circle
A pasture of swine dying of cholera—
Definite rounds, awful solitude

Of meat going bad for good:
The women bustling at the bench
Withdraw fire the washpot keeps faint.

II

My childhood gets lost in rituals.
The hogkilling day hears more knocks
Than my brain can assimilate,

Cutting didoes in the thaw of Paul's Hill,
Kicking off noises of motor-cars
And the shine of crosses and rusty nails.

The white sheets hover. I need them
To sleep on, far from bullets and knives,
The gambrels, blood, whiskey, chatter,

Prevalence of witnesses guffawing,
Talking through Tube Rose Snuff and Beech-Nut.
Presences time chores.

III

Now strike the scene as gold
For the pot, the cracklins smelling bold
As meat between a duck's foot the sun

Lifts over salt-pork boxes
Turning eye-level devilment
Of booze and desire foraging the crop

Which gets laid by once only,
Pulling back the tobacco canvas,
The undermine of gosling weight

The head of the house, a tenant,
Fixes in torchlight of gaggle,
Blazes among litard-knots.

IV

Wharf-rat the texture of fuzz-wear,
Underpants gray with farts
Spreading rancid shake and shine

Of one thousand songs,
Plus news of rounders getting cut on weekends,
The likelihood of absence the day of the kill,

The grin of heightened awareness
Of ones who do show, the graying
Rats scattering the fire's blaze

Which cuts the crusts of thaw in the outhouse.
I never saw two people in it. And it was
A two-holer: we called it The Backhouse.

V

If you see me, darling, in some pig-skin suit,
Say I love the ribs of thighs and memory
Of the pants my shoat bloodied.

Tell the gloom to lift so I can hear her.
My wrists twist in squeals of braceleted
Giddiness for love of my Duroc.

My pores race
Follicles to my elbows as far
As I can see to break up moments

We shared together, especially the time
We walked hand in hand to the chapel
In the woods to see the pink pigs.

VI

Now the goddess of entrails comes to work,
The head-woman passing about the guts circling
Humanity round, marking ways to sling

Up and down and side to side, intestines,
Not showing off—this was not a fair—just freedom
Courting stink-doom, welcoming

Mornings of chitlin-cleaning
Rituals while the women pause in slow and strong
Murmurs of malodorous guts—

How good they will taste with collards and ham-hock
Once the work is done; the women will wait
To put up preserves and pickle some feet.

VII

Spring memory's smell and fabric of air,
A colony of well-clothed women in cut-off socks
Displaying their arms, structures: they

Learn to feel the grainy casings
They use after the peach-twig turns
And courses the water to pour

And scour—scald the insides swaying and turning
Wrongsideout into memory's setting
Forever blank; a portent retches *Go*

With little silver lights all in a row.
The pigs that were free for a while roam
The pasture where shadows play.

VIII

You can't say "guns" much anymore—
Though there was a time I could take my
Twelve-gauge Iver Johnson Full Choke

Squirrel gun unbreached across my arms
And walk out to Beaver Dam and gloat,
Just get simply turned on from being

In the woods, sensing the eternal blight
Of longing, the sin-arrow missing the mark,
Whatever small game I shot for the table,

For the galaxy of things to appear there,
Seeing my father's face before me, his smell
Of hunting pants and jacket—and my mother's eyes.

The Story's the Way

Let us consider the mess of chitlins on
Maytle's table, plus the past, for me—
A history which conjures Pap George's own
Slaves: he thought he could keep them "property,"
As was the law, a tainted one ever
Enlivening the Chitlin Circuit—
Juke-joints, clubs, keeping African-American stars
In love with their artistry.

I was going to say
I should prefer some woman at the table,
An odor, faint, of innards in action,
Since women got the chitlins
Ready out of doors instead of in the house.

The whole yard is a hot stink. The oak trees
On Paul's Hill stir grumblingly
Inside that smell, an envelope
Of tripe, souse, and haggis, memories
Instinct locks in your body: you face
Images of Moms Mabley during the Depression
Or Little Richard rocking and rolling like a dolphin
In a circuit you cannot disconnect.

Satch scatters, swipes his mouth, chants one-two-three,
Scanning history the way Mama Maytle
Arranges the washtubs as if on a bed,
One washbench under her chainey-berry-tree
Where she washes and wrings our clothes by hand as
Ivy rises with guts she cooks on her Home Comfort Range.

It Is As If

It is as if I have
nothing to say really;
refrain: Paul Junior's gone,
my sister Maytle Rose,
too, my father, the grand
Paul *S R*, my Maytle,
his wife, still coming back
almost into these words
merrily to show some
feel for my birth-house
hunger to be reborn
to tell one thing you might
sing after the skipper
flies my mother cut out
of the ham, though she said,
"This ham will always taste
tainted." My heart wobbles
for a kiss to come and
unhang her father from
the rope he tied around
that mulberry limb near
St. Mary's Church and jumped,
her twelve-year old brother
seeing this, still running
across the tobacco
bed near the hog-parlor.

Confederate Soldier

(for Great Grandfather Manly, died 1912)

They say he walked home from The War and quit
The kind of life a plantation can fit

Into history, especially this
Century's emphasis on revision.

Confession: when I write "Confederate"
I am holding back to soften the fact

That I am his great-grandson, looking on
His confederate musket carved in stone.

I salute artificiality
Lying there. And the slaves? Formality?

Great-grandpa Manly's dad, Pap George,
Farmed seventeen who made land and crops forge

Upright and prosperous, plus his two wives
And nineteen children. What unsettling lives

His farm must have required to work thousands
Of acres of land and plenty of sows

And pigs caught in a world history made
By our forefathers and mothers who laid

Bare the legality that caused the spirit
To seek new silk for old sacks to clear it—

The separate but equal, black and white
Or brown, I mean, to savor Love's surfeit.

The True Story of Chitterlings

Smelling up everything

It mouths
Like a vowel
Over real bowels.

The Orchard Boy

I

Now one time I had an apple orchard
And time was apple blossom white
  With a heaven made in treecrotches.
    The mules spraddled their legs, blinked pink
     aftermaths of pissing.
  Roadapples I stored for bases in the yard.

Across the dirtroad the garden lay,
Paradise of okra, peanuts, gardenpeas, squash, tomatoes,
  Maytle Samantha's slip falling below her hem,
    Cucumbers, potatoes, radishes, turnips,
     butterbeans, collards,
  My bare feet hopping clods.

The hedges were handsome as the doves
I stalked with my Daisy:
  Hi yo Silver
    Saddled me
  And we rode from gummy, tobaccoblooming fields

The ten miles to town—
Warehouses—the smell of cured tobacco
  I chewed in those suncured days
    The Neuse River rippled with shad,
  The farmers on the banks working their seines.

My seat, a burlap sheet torn in two—Middle Creek!—
Leafboats sank in puddles kicking round
Beaver Dam where flapping tails went silly
And teeth
Gnawed tulip poplars down.

Thicksmellingrich with cornshucking and fall's fodder
I stripped and tied in bundles
Shocks beside haystacks,
The caves underneath where I jumped a rabbit
And sat to watch for the orchard boy.

II

Blue Boy, your sheep grazing wirebarbs,
Your haystack, a mouse's trove
Birdstartles sweep into the meadow's wad of cows
Along the lake, bluewaved,
Owls whooing darkness,
Dear sleep's drool,
Pillowing wakes.
The bear licks gold porridge
The highchair swamps.
The valley
The days tally,
Locks—golden braids—
Breezes turning heads,
Catching
Sleights of trees
Leaftongues release.

III

Birdflights
Out of rhyme's
Sweet time,
Dear ones, remember
The sheep grazing the fence where screams
Stopped inside your hour.
Faith muddles
The thorny nook.

Bring humors to bear.
Ring bells for girls
And boys with lyres.
The pole that stood
Tall on the schoolground yesterday, children
whoop and holler about,
Now: round and round their shoes dig dirt,
Their faces, fiery, muffled, shout.

IV

Spinning off a platter
The boy's prize porker
Pedals toward the Animal Fair,
A dancing arm stretched across the handlebars,
Chin propped on her right front hoof.

V

It is a roundabout sleight
January whiffs over hogs,
Flatbedtight,
The boy's nose a running sight.
Bloodkill
Dangles over dogs
Howling for a morsel
Dripknifecarves.

The Wind in the Woes

for Fred Chappell

The scene is the cradle or the grave.

The possum turns its soft belly toward the moon. The persimmon tree sways the path. The field's a field, but the way the light falls, shadows perform an act of great closeness.

It is all the wind can do to be still. "Let the characters out," it says. Cloud laughs, bemoaning the light dillydallying on the ground. Other voices play around—there is Cat, the Old Woman, and Shoe Tongue. Certainly a new world is at hand.

The horses wallow in their stalls and paradoxes. Oxymorons scratch their heads. The monkeys crack their peanuts and turn away.

A rider on an oldBaptist mule trots into the churchyard and ties the reins to a wirefencepicnic table. A Lumberton woman rides her pig over to a tombstone and the pig, a Duroc, snuggles next to the cool slab.

No one knows what's going on. A sense of drama seems inviting, but nothing happens. This is farm country and you can see the enchantment and the hope that the characters will come and make the crops, but all they want to do is play.

Horatio Humpty lies against his pet pig's belly. He has dressed the shoat in a KKK outfit. In real life the pig is a racecar driver; the skit presents him as a porker bent on getting as much as he can.

Mary Quite tends her bells. Buster Brown brushes his blue suit and winces. Tige chases its tail. The fleas scatter, looking for a home, unwelcome in such a lonely conclusion.

Wind

We have gathered here to seek the dead. In the quink
of an eye God takes away those He loves.

Tree

My characters reside in me happily. Sylph is my
unshifting loyalty. My leaves die, but remember April when
Passion is my lover and I am all spread. The whole question is
whether to branch out and go beyond the hole in the fence.

Wisteria Chapel

Whose daydreams in yonder forest
Usurp wisteria's soulful
Purpling among the gentle looks
Of pigs suckling a sow's teats?
Should I as witness be so bold
That peering takes me from my books?

The pink piglets need not my rave.
I can't taint the scene with remorse
If the mother's breathing pants plaints
Which beg me go and no more crave
Exposure pondering choices.
Bed's a needled carpet for saints.

Yet she is my 4-H project.
Her name's really Shelby's Lady.
Pedigree, written big, says so.
There, I've said her title, smoothed out
The matter; conscience, let me be.
My father says, "Just a hog, boy."

Father, her eye-lashes, closed, pierce
Into a stony block my heart
Softens to loosen her labor
Subtle amid vines she insists,
Without orator, tones sparkle
Duroc red winning my favor.

What Is Your Sign?

Vary the boring stricture arrears: forget the bills. Give me mugs,
shirts to advertise the horoscope, a random word or three to
send Woodstock yellowing when
Snoopy snores. I want to cuddle and bug Love's chassis; dodge
oil and water in filling-station parking lots. Give me Castor
and Pollux; Virtue, shine. I am a sun of constancy.

Fickleness far from swoon, I appreciate, too deep and wide to
hold you and me both in a hammock. My mother and my
sister could really cook ham-hocks. My brother Paul's air

and dust: possums mooch his grave under a dogwood.

The last of the Bug Eyes, my Sprite I sold for honeymoon
money. O hot gods of heart no sun can delete, yank a space
void of blame and show me your mild side, for my mother,

Maytle Samantha, and for Leda of wrestling fame. Let my
name, Shelby Dean, roll along, twine Cancer and Taurus.

Wave The Flag on June 14, when I was born, 1938, 161 years
after the Second Continental Congress by resolution
adopted the flag on Flag Day, the same day the United
States

Army celebrates Army Day. Salute 1949: National Flag Day
established. In Troy, New York,
a Flag Parade annually draws a lot of people.

My mother, Maytle, and my father, Paul, did not know the word "condom" any more than "condominium"; honor was their bond as honorarium's sign to Gemini which runs May 21 to

June 20: artistic, nice, intellectual, individualistic, outgoing, compassionate, plus the opposites; on the other hand, a golden band would make me nervous, inconsistent, dotey.

Finale

1

I have to cross back through the pig's eye,
Where the sty in my eyes
Greens hollies at woods-edge

And sings songs I can't forget—how the knives
Seem to hone themselves and the men's
Overalls claim a fading blue,

Uncle Walter's pickup parked ready
To roll back to Holt's Lake, his whiskey
A secret he keeps to himself

In the out-buildings where he stores his bottle,
His wife, my aunt Ruby, mumbling
Constantly the way an independent woman copes.

2

The truth lasting, in memory on this hill,
Translates this day, Memorial Day, 2019.
Into the squat military lots in beds of trucks

Going toward Fort Bragg in a convoy,
Letting every boy and girl by roadside
Twirl time on bicycles, yammering warnings

The stilled jaws of clenched turnings,
A country music we all learn,
Leaning and yearning for peace,

When no attention gets paid enough to meat,
Just the hogs and boys at feeding time,
The slow life born on earth's smell of heaven.

3

The morning flows out of these sticks to scatter
My mind in a million snippets: the feeding
Ground the hogs made their own

So that I must feel like an intruder, trucking
Around, my eyes tagged with sleep of
Early morning, the rounds on the little

And big John Deere tractors,
The cargo of brightness turning on
The wriggle of one pigtail

Lil Sis swirls in her grip on a shoat
She's got by the ears in the pieces
The sun holds once only in blues.

4

The piglet was like a baby. We lined up
The nipple too snug for much
More than an oink and grunt,

My hand scratching its belly of pinkness
Into a motion its own award,
A farewell to future ribbons at the fair,

The best for my baby and for me.
The rest I do not remember, except one day
As killing-time approached, my father

Thought, *You are mine and the hog's mine.*
I could smell the anxious smatterings,
Like several sparrow-hawks looking for a home.

5

Scissors-and-hums: motions of an angel.
Her fingers and thumbs. The patterns lift
And drop in breezes hemmed without harshness.

This is a mother timed to music and living
Off the land, spring, the meat
Curing on the poles of sweetgum nailed to the ceiling

Of the packhouse; there she is, moving around
The eatin-table in a threshold of possibilities,
A known world, yet adjusting

Her hands, heart, the fanning heat,
Impulse, willfulness, the need to get by,
On what she knows she can.

6

Sometimes she'd find skipper-flies, she called them,
Drunkards, eating at a ham hung to the poles
And she'd take my father's handsaw and cut

Out the tainted part; we ate the ham,
Though it never tasted right after that.
The open door smelled of usefulness every day

And kept us fed: our heads were free of lice,
Though I heard Harvey Coats got them in the fourth grade.
I was borne into gleanings of my new corduroy coat.

Harvey used his sleeve as handkerchief
And one day on the playground he held one nose-hole
And let fly a greenish praying mantis of a stream.

7

Not all cars in the '50s reminded me of hogs.
Some did: the Kaiser, Frazer, Henry J.
And one time Uncle Dave Johnson (he called his violin

A "leedle feedle") drove up in a powder-puff
Blue Henry J. A bull-fiddle was strapped to the top.
We drove to Meadow in that thing and played

A talent-contest—and won. I sang
A song Hank Williams did not write: "Window Shopping."
The parking lot at the schoolhouse

Was filled with Hudsons and Nashes and shapes
Of Poland Chinas, I thought, Durocs.
We won $25.00 for the prize. Music may be the
    poetry of the soul.

8

I never missed running water until we got it.
I won't forget the well-rope in my hands,
My arm-muscles making frogs mid-arm.

The water-bucket moistened the ledge. I'd tilt it
At first and hear the topple level off into flatness.
Cold water. Deepest well in the neighborhood, my father said.

Probably was. It's all in the head. The well
Was on Paul's Hill up from the hog-parlor
Down below where the lot-well was.

It had a chain. My hands still hurt from the knife-sharp
Blades working my body, never to cast
Off and span what memory cannot forget.

9

I can almost tell you how *it* was. Recollect
Driving to his room at Rex Hospital,
Where he was already dead.

The nurse said he looked at her
And rolled his eyes and said, "Don't you know I'm dying?"
I think about that. Then the wind comes up

And I forget where I am, leaving to drive
To my warm bed and leave my father alone,
Never to stand more on firm ground

Or rassle with the boys during fodder-pulling time,
Never to hit a hog again in the forehead with a maul
Or shoot a big old boar between the eyes with a .22.

10

MacLeish said something I can't get exactly right:
In his "Memorial Rain," time changes the skin
And dries his brother's self to rise more into what

Valor might be, his face, a baby's, down in mud
On a field left for the birds, unless someone finds
The newly dead to recognize a coming back,

Family glad, grieving, distraught, their soldier-boy,
Found, their farmboy who wanted to scratch his pet pig's back,
And feed him slops and dimpling swill from the bucket

In the kitchen, far away, his other worldly
Head, lonely, low, his beard, the not-to-be surprised
Responses family orders across the sea.

11

And so I come to you, my piggledly-wiggledly
Golden bower-keeper, my pretty wolly-doozle all day,
My blinking cursor, my coursing sentry,

My Miss Piggy-waddle, tail-twitching bun-in-a-blanket,
With a pig-in-waiting, a sow of a century just over,
Though one never is, memory being a show

Of a different color, a dolor of sunrise which reminds
Me of you, your hair, your brows, dark as a treeline
At dawn, a yawning coup of bristles,

Pickled pigs-feet, eggs in vinegar, and you,
All over me, my Love in Happy Valley, my ears
Unstopped in the sun, our lobes over the dancing floor.

Insert Coin Here

If you could put a coin in my video, you would see an eighty-
one year-old freckled man, born under a collard-leaf
shaped like a three-room house, my mother, thirty-three-
years-old, looking not a bit like Mary. My father said I was
gotten from not pulling out soon enough.

If images play by ear, leaves convert impulses from saving sweet
sounds of popping rain on a tin-roof my baby-self swims
in: how comfortable I am, riffing frames to qualify my life,
compressions no technology may stream, for my mother
wanted a girl: here I am.

Shelby Dean Stephenson—instead of Shelby Jean (for Davis),
my namesake, The Little Mountain Sweetheart, WLS,
Chicago, Friday Night Frolics. "Dean's" for "Dizzy" (my
father loved baseball): if I were a video I'd be SDS instead of
DVD; I'd be Red-shub instead of

Blu-ray Disc searching to send pictures in clips. Truth is I don't
want to merge in other words. My lover never lets me go;
living's a vision for high definition. Consider pixels which
cannot possibly get fat on a steady diet of collards and hog-
jowl, plus pickled pigs-feet I grew up on.

I know a hog when I see one. I wish I could put another
nickel in the jukebox and let Rudolph Wurlitzer rock
with Shelby's Lady, my pet Duroc, or Shel, himself,
making speaking contests in Future Farmers of America.
The bite would show Shel talking into the wrong end of

the telephone first time he ever talked on one when
Mr. N. G. Woodlief, his principal, called him to his office
to take the call.

The Ending

And there in the boat which came toward me,
My brother, Paul, a cinder-block in his arms
And he threw that thing (it was hooked to a chain)

Forward as if to anchor his craft never
To return to weeds and land again,
His eyes and bald head a rolling tumble

The splash pulled on and in until
He stood in mud (the pond was not that deep)
And I thought, "That's the kind of person he is."

I'll bet a song was in his head,
For he was the loveliest of brothers,
Ten years older than I, quiet, until, say,

A hog might run for him in the marsh,
If he got too close to her piglets.
Whose hooves these are the boat never

Needed, wheels of carts and golf-chariots,
Spare tires, tar and nicotine, pig-ears,
And balls, big old boar hog balls,

I mean, ones which need cutting out
With a Gillette Razor Blade so that I may
Conjure Mr. Roof Allen, one-eyed

Roof, one more time, while his wife
Thelma waits for me to come and sit by her side
While she hums and makes me a monkey

Out of a worn-out worksock,
One with that heel darned and dyed with red,
To make the monkey's ass shine

Straightaway on the road which winds
From some Eden to Now, the couple, first one,
Not squalling, but wanting to, holding back

The crying until they can get turned
On to each other's words, their hard bodies
Softening as they walk naked and exhausted

With worry over how to cover their privacy,
As we worry this Sunday morning how to dress
In our Sunday best for church where the wheels

Of fire turn for me once more,
Hearing the preachers of my youth
Spin the prayer-wheel,

Thump and parade, praise the chattering
Lord God Almighty, bring on the teeth and the jowls,
Souls, lost at dawn, those needing

To be borne out of living's forms
Where music resides deep and wide
Right down in the fanning rhythms syllables

Lay down like bass bedding full moon in May,
Crowding fins and fattening
Little eggs hatching unceremoniously.

After Words

In snouts of snores I come to you,
Sink up in your belly of lard.
You grunt the score and I come to
Like pickled pigs-feet Mason-jarred.

Title Index

A

B

C

D

F

G

H

I

J

N

O

P

R

S

T

First Line Index

A

B

B

D

G

H

I

L

M

N

O

P

R

S

T